CW00460554

THE LEGACY OF ALFRED NOBEL

THE LEGACY OF ALFRED NOBEL

The Story Behind the Nobel Prizes

RAGNAR SOHLMAN

Translated from the Swedish by
Elspeth Harley Schubert

With a Foreword by
Stig Ramel
Executive Director of the Nobel Foundation

THE BODLEY HEAD
LONDON SYDNEY
TORONTO

British Library Cataloguing
in Publication Data
Sohlman, Ragnar
The legacy of Alfred Nobel.
1. Nobelstiftelsen – History
I. Title II. Ett testamente. *English*
068.485 AS911
ISBN 0-370-30990-1

Originally published in Swedish
as *Ett Testamente* 1950

Printed in Great Britain for
The Bodley Head Ltd
9 Bow Street, London WC2E 7AL
by Redwood Burn Ltd, Trowbridge
Set in Linotron Plantin
by Rowland Phototypesetting Ltd
Bury St Edmunds, Suffolk
First published in Great Britain 1983

CONTENTS

FOREWORD

On 27 November 1895, Alfred Nobel signed his last will and testament in Paris. Among its four closely-written pages, less than one referred to the donation which was destined to link his name with the supreme achievements of the modern world in science and literature, and with the cause of peace. Nobel's conception of five prizes 'to those who during the past year have done humanity the greatest service' was ingenious, and he bequeathed one of the largest fortunes of his century to its fulfilment.

The will, however, was found to be legally deficient, and contained flaws which soon aroused criticism and protest. The testator had given no indication of his plan to the institutions designated to award the prizes. When it was opened after his death a year later, its provisions came as a shock to his family, and gave rise to doubt and misgiving among the proposed prize-givers. The aim of the legacy might well have remained a brilliant but impossible idea, doomed to founder in a Sargasso Sea of conflicting opinion.

The fact that Nobel's visionary dream came true—in a manner which he himself could never have envisaged—was largely due to the appointment of his young secretary and assistant Ragnar Sohlman as one of the executors of his will. The conferring of such a responsibility on a twenty-five-year-old—a daring venture at any time—might have been disastrous, but proved instead to be the result of shrewd foresight.

Looking back on the dramatic years 1897–1900, Ragnar Sohlman's handling of his task calls for the greatest admiration and respect. A chemical engineer like Nobel himself, he was employed by the latter in 1893 as his personal assistant and, during the short time they worked together, won Nobel's wholehearted confidence and affection. Sohlman was a small man, modest and retiring in his manner, and certainly not, to outward appearance, cast in a heroic mould —but it soon became evident that he had plenty of fighting spirit and was fully able to cope with the immense difficulties which threatened to shatter Nobel's intentions.

After three years of difficult negotiation and continual setbacks, Sohlman, his co-executor Rudolf Lilljequist and their legal adviser Carl Lindhagen managed to attain three main objectives, namely:

1) the legal recognition of the will, and its approval by the Nobel family;

2) the realization, in accordance with his directives, of Nobel's assets and their placing in a fund, later to be known as the Nobel Foundation;

3) the agreement of thc proposed prize-givers, i.e. the Academy of Sciences, the Caroline Institute, the Swedish Academy and the Norwegian Parliament, to accept their assignment.

The first task was the most difficult and delicate. Nobel's family bitterly resented what they felt to be the inadequacy of their inheritance and there were grave fears that the sale of Nobel's shareholdings both in the big Russian oil company, The Brothers Nobel and in the various dynamite enterprises might threaten their financial position. However, the branch of the family resident in Russia and headed by Nobel's nephew, Emanuel Nobel, soon came to an agreement with Sohlman, and Emanuel himself supported the executors in the implementation of the will. A very wealthy man, who had held his uncle in high regard, it was he who reminded the young engineer of the Russian proverb which calls an executor *Dusje Prikastjik*—the vicar of the soul —telling him to live up to it. Sohlman came to feel that he embodied Nobel's last wishes, and this gave him the courage and determination to carry out the duties that lay ahead.

These included extensive negotiations on behalf of the estate with contemporary leaders of finance and industry around the world; resistance to opposition from every quarter, not least from King Oscar II and the leader of the Social-Democratic party, Hjalmar Branting; skilful manoeuvring between Swedish and Norwegian interests when the final dissolution of the Union became imminent, and continual journeys to the different countries in Europe where Nobel's own interests were involved. A dramatic incident was the transport of valuable securities through the streets of Paris in a horse-drawn cab, with Sohlman, a revolver at the ready, literally sitting on Nobel's millions to protect them from being stolen! In the midst of all this, he was called up in 1898 for military service in an infantry regiment in the centre of Sweden. As Recruit 114 Sohlman was probably the most remarkable Swedish conscript ever—his telephone calls and telegrams to various European capitals, and especially to antagonists of the Union in the Norwegian Parliament in Oslo, surprised and alarmed the local military staff.

In the summer of 1898, a legal compromise was reached with the Swedish branch of the Nobel family, which allowed them certain economic concessions, and also contained an agreement regarding the distribution of the Nobel prizes. The struggle which had engaged many of Europe's most skilful lawyers had been long and hard—but these wounds healed quickly, and today the Nobel family is proud to see its name associated with the world's most illustrious awards.

When, in 1900, Ragnar Sohlman was finally able to hand over the administration of the estate to the newly-created Nobel Foundation, he was admittedly elected as a member of the board, but received the lowest number of votes among those chosen by the prize-giving institutions. The differences of the past years had left their traces . . .

Sohlman now became managing director in Bofors Ballistite, and continued to live at Björkborn, Nobel's Swedish home, from where he had led the battle over the will. In 1919 he joined the Board of Trade, finally becoming its head.

From 1936 to 1946 he was managing director of the Nobel Foundation.

Sohlman died in 1948. His last years were spent in recording his personal memories of Nobel, in particular the period between the latter's death and the year 1900, when victory was won and the Nobel Foundation instituted. His book, under the Swedish title of *Ett Testamente*, was published posthumously in 1950, and aroused great interest, not least for its account of Nobel's strange love story with Sofie Hess, a subject which— in view of the discretion of the time in such matters—had not earlier been enlarged on. Much of the documentation about Nobel is based on Sohlman's book. As the only biography by someone who knew him personally, it gives an authentic picture of his way of life and the circumstances of his will. The book now published contains the full text of Nobel's last will and testament. This is the first time it has been made public—only the part dealing with the five prizes has previously been made available. Eighteen people, comprising members of the family, friends and servants were remembered with very substantial sums of money. Nobel was as generous in his will as he had been during his lifetime.

The fact that the book has not been translated before into English is an omission which the Nobel Foundation wishes to repair in connection with the 150th anniversary of his birth in 1833. We hope that this will help to spread a wider knowledge abroad of Nobel himself, his work, and the story behind the prizes.

The updating and editing of Sohlman's somewhat antiquated text for an international public has been entrusted by the Nobel Foundation to Elspeth Harley Schubert. Certain sections at the end of the book dealing with lawsuits, patent rights and detailed negotiation with the prize-giving institutions have been deleted. The Nobel Foundation would like to express its appreciation and gratitude to Mrs Schubert for her handling of this difficult task. A contribution towards the cost of the translation has been made by the Anglo–Swedish Literary Foundation in London, which was established by a donation from George Bernard Shaw from his Nobel Prize for Literature.

Dr G. H. McCallum of the Nobel's Explosives Company kindly checked the chemical terminology. Mrs Margaretha Ehren gave invaluable assistance with proof-reading. Mr G. P. Bartholomew compiled the index. Finally, the Foundation wishes to thank Dr Ragnar Svanstrom for his interest and valuable advice.

The publication of *The Legacy of Alfred Nobel* in English is not only intended as a tribute to the memory of the great donor, but also to that of Ragnar Sohlman, without whose faithful and untiring efforts the Nobel Foundation would probably never have been formed, and the idea of the prizes only remembered as a flight of fancy in a controversial will.

Stig Ramel
Executive Director of The Nobel Foundation

NOTE

At the time of his death Alfred Nobel's fortune amounted to Sw. Kr. 31,000,000 or £1,706,100. This is equivalent to Sw. Kr. 692,000,000 or £60,000,000 in present day currency.

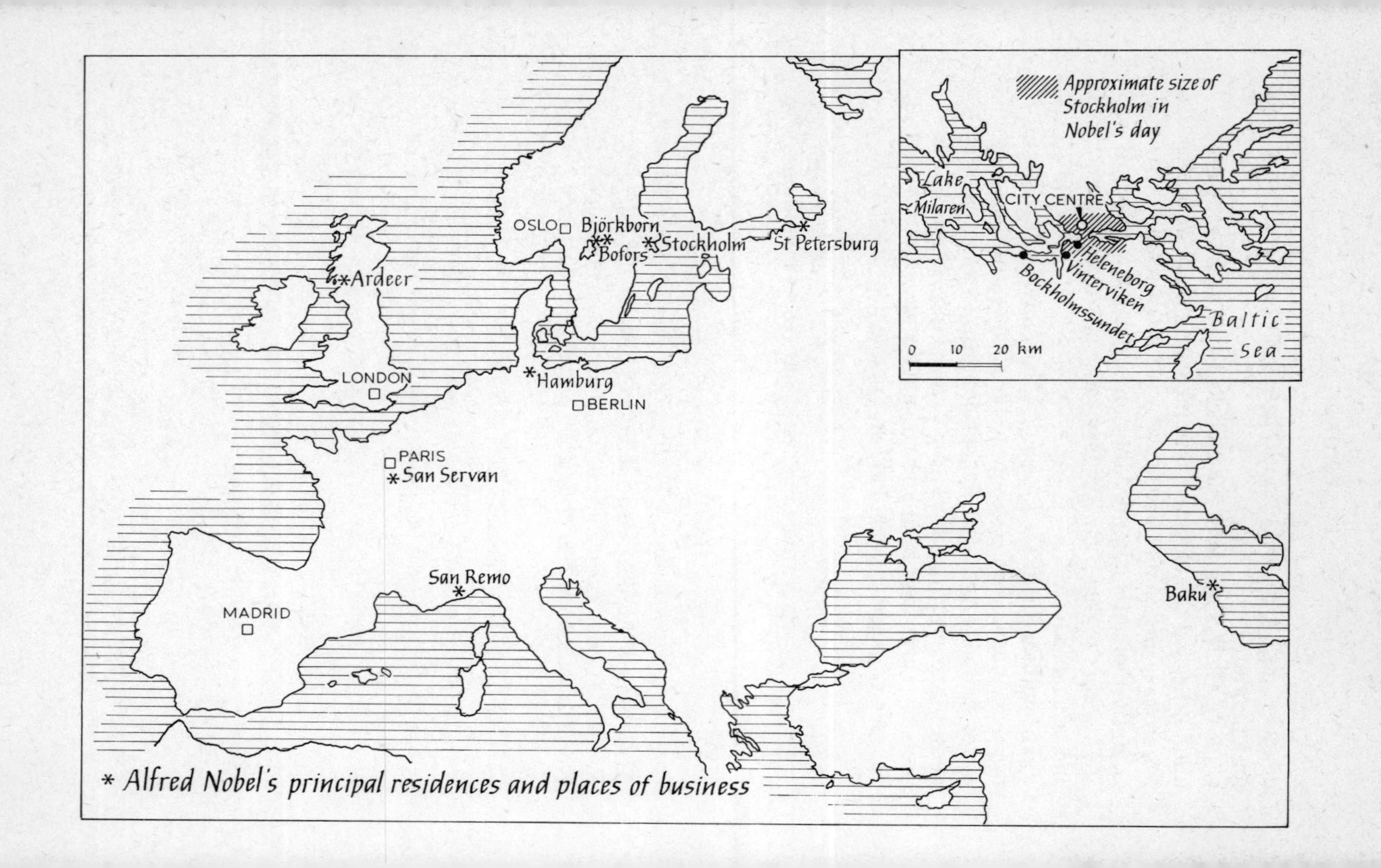
Approximate size of Stockholm in Nobel's day
Lake Milaren
CITY CENTRE
Heleneborg
Vinterviken
Bockholmssundet
Baltic Sea
0 10 20 km
OSLO
Björkborn
Bofors
Stockholm
St Petersburg
Ardeer
LONDON
Hamburg
BERLIN
PARIS
San Servan
San Remo
MADRID
Baku
* Alfred Nobel's principal residences and places of business

INTRODUCTION

It is now more than half a century since Alfred Nobel's many-faceted life came to an end. He died in San Remo on 10 December 1896, at the age of sixty-three.

The publication of his last will and testament in early January 1897 caused surprise both in Sweden and elsewhere. In this document he left the bulk of his estate to form a fund, with the interest to be distributed annually 'as a reward to those who during the past year have done humanity the greatest service'.

The terms of the will, its unusual purpose and partially incomplete form created a sensation. There were many sceptics, and criticism was soon levelled at the donor for his 'lack of patriotism'. Only after four years of investigation and often bitter argument, and when objections of every kind had finally been overcome, could the will be given definite form and the Nobel Foundation be constituted.

Through a Royal Ordinance of 2 June 1900, the statutes of the Foundation were established, as well as the special regulations governing the Swedish prize-giving committees.

Now that over fifty years have passed since the first Nobel prizes were awarded, criticism has subsided. The prize-giving has become an annual event of international importance, and Sweden's prerogative in choosing winners in the fields of physics and chemistry, medicine and literature—an exception being the peace prize—is considered an asset, enhancing Sweden's reputation as a cultural centre. Meanwhile, interest in Alfred Nobel's personality has increased over the years, both at home and abroad.

Nobel's contemporaries and close friends are all dead, and

only a few, born later, who came into contact with the great inventor and idealist and could contribute to a personal picture of him, remain alive.

A comprehensive and sympathetic biography of Nobel by Professor Henrik Schück was published in 1926 by the Nobel Foundation, to which I myself added a survey of his work and life story. For reasons which I shall explain, I now feel able to give a more intimate description of his personality and to interpret the origin and intention of his will and its conditions.

As Nobel's private assistant during the last three years of his life, I was in almost daily contact with him, either personally or through correspondence, most of which dealt with the experimental adaptation of his inventive ideas. A clause in his will of which I was not aware until after his death appointed me one of the two executors whose charge it was to wind up his estate and take the necessary steps to implement his final instructions. It was clearly his intention that I should shoulder most of the work—a responsibility for which, considering my age and inexperience of legal and economic questions, I was poorly qualified. There was nothing for it, however, but to go ahead—and as a result I was obliged to take an active part, not only in the liquidation of his financial interests in various countries, with all the complications this entailed, but also in the arguments about the validity of the will itself when this was disputed by some of his relatives. In addition, I had to join in lengthy discussions with representatives for the appointed prize-givers, and attend committee meetings to establish the basic statutes of the Foundation. Then followed the task of sifting through Nobel's immense correspondence, recorded and copied in folders, and of studying his personal files. These, of course, included private letters not previously examined, out of respect for people still living. All this has given me a clearer understanding of the circumstances which influenced Nobel in the drawing up of his will, as well as an entirely fresh view of his mentality and emotional life.

Ragnar Sohlman, Stockholm, 1947

Personal Memories of Alfred Nobel

The name of Nobel had been familiar to me since childhood. While busy with social and cultural work in Stockholm at the beginning of the 1860s, my mother had made friends with an energetic lady named Mrs Elde, who was a sister of Immanuel Nobel, and thus Alfred Nobel's aunt. At the time, Immanuel and Alfred were experimenting with nitroglycerine in their tiny laboratory at Heleneborg, near Stockholm. On 3 September 1864 these experiments were interrupted by a violent explosion in which five people were killed, among them Alfred's brother Emil, a boy of only twenty. This was a crushing blow to his sixty-year-old father, who was working with them. He had a stroke soon afterwards and, for the last eight years of his life, was not capable of any serious collaboration. This meant that the entire responsibility for the development of new ideas in the technique of explosives fell on Alfred alone. Like his father, he had not the means for any large-scale manufacture of his inventions, and was also threatened with heavy damages. After the accident it was not easy to induce people to put money into what seemed such a risky business.

Through my mother and Mrs Elde, Alfred Nobel was put in touch with a cousin, a successful businessman named J. W. Smitt, whom they managed to interest in the Nobel inventions. Smitt later put money into the newly-formed Niko Company—probably the best investment he ever made. As a child I used to hear Smitt refer to those early days and his visits to the barge at Bockholmssund (near Stockholm) where the highly primitive manufacture of nitroglycerine was carried on, amid a haze of cigar and pipe smoke, and the finished product, blasting oil, transported—often by farm lads, some of them drunk—to Stockholm.

It may well have been these reminiscences which made me decide to be an explosives technician. While I was still at school I spent two summers working at Vinterviken (Winter Bay) the oldest Nobel explosives plant, as an apprentice. Nobel's lightning visits from abroad to his homeland and the plant were discussed there with due respect. They were nearly always connected with new inventions and improvements in the technique of explosives, most recently the discovery of ballistite, which the Swedes christened 'Nobel gunpowder'.

One summer, while I was studying at the Institute of Technology in Stockholm, I took the chance of a trip in a Swedish tanker which was carrying naphtha to the Caspian Sea. These journeys were made during the spring tides, by river and canal via Ladoga, the Svir River and the Marinsky Canal to the Volga, the Caspian Sea and Baku. In Baku I went to see the Nobel plant and drilling field and in 'Villa Petrolea' met Robert Nobel, whom I knew already through one of his sons, with whom I had been at school. From Baku I embarked on a long trek by myself, into the Caucasus. To escape notice I wore Georgian dress with a *kinscal*, and carried a gun. Even so, I was often asked where I came from, and this was difficult to explain, since the word *schved* —Swede—meant nothing to the country people. However, as soon as I mentioned that I came from the same country as Nobel, the comment was always the same: 'Oh, you are *Nobelskij*—a Nobel man.' To simplify matters, I finally adopted this title—little thinking that I was to become a *Nobelskij* for most of my life.

After graduating in Stockholm, I went to the United States between 1890–93 as an engineer, working part time as a chemist at a dynamite plant belonging to the Hercules Power Co., nowadays a subsidiary company to the famous du Pont concern. In the summer of 1893 I helped to run the Swedish department at the World Exhibition in Chicago.

When this closed down, I was offered a job as engineer at a projected dynamite plant in Mexico, but since my health was not too good at the time after an attack of pleurisy my family urged me to go back to Sweden, or at least to Europe. I had no idea then where or when I should find other work.

My surprise and delight can be imagined when in September 1893, while I was still in Chicago, a cable came from Stockholm to say that I had been offered the post of private assistant to Alfred Nobel, with immediate effect. It turned out that J. W. Smitt had recommended me. My friend Ludvig Nobel, Robert's son, had also put in a good word for me, although I did not know about this until many years later.

Needless to say, I accepted this tempting offer with alacrity.

A week or two later, admittedly feeling a little nervous, I went to present myself to the famous inventor at his private home in the Avenue Malakoff in Paris. A butler showed me into Nobel's study, where he received me with the utmost friendliness.

Alfred Nobel was then about sixty years old, slightly under medium height, with strongly-marked features, a high forehead, bushy eyebrows and deep-set penetrating eyes, as lively as his temperament.

My first task was to rearrange and catalogue the reference library and files he had brought with him to Paris. The library was fairly large and included scientific and technical works as well as literature in French, German, Swedish and Russian. Later on, Nobel made a comprehensive collection of Norwegian and Danish classical authors—Björnson, Ibsen, Lie, Kielland, and such Danes as Oehlenschläger and H. C. Andersen.

To start with, I arranged the books, which were in some confusion, putting the technical and literary works in separate sections on the shelves. This took me a day or two and seemed to please Nobel, making it easier for him to find what he wanted quickly. The many personal and technical files which had been placed in envelopes with Nobel's signature and a list of their contents took me two weeks to sort out, working at top pressure. I stayed at a small hotel nearby, but lunched every day with Nobel in his conservatory on the top floor of the house, where I was the only listener to his interesting and witty conversation.

Whether he had expected to find me a suitable secretary from the start I never knew. It must have been clear to him at once that I was not sufficiently qualified, but if he felt any disappointment he never showed it. He himself was a skilful linguist, able to write fluently in French, English, German and Swedish—in that order of proficiency—and also in Russian. The fact remains that he never found a private secretary whose style and proficiency in languages came up to his expectations. He had, too, a certain instinctive reluctance to handing over the personal care of his voluminous correspondence to a stranger.

In a letter to a friend from San Remo, dated 24 May 1893, Nobel wrote:

> . . . I am not lazy about writing, but tired of it, and now have to think, *coûte que coûte*, of engaging a male or female secretary. Preferably one of the latter, since they are less troublesome when one can't keep them fully occupied, and can usually find themselves an admirer, so that one doesn't have to trot them about in the same way as a chap. . .

Reading this letter long after Nobel's death made me feel slightly conscience-stricken, even though there was never any question of his having to 'trot me about'. His search for a woman secretary resulted in his taking on a certain Miss Sophie Ahlström, to whom he wrote explaining his requirements:

> For me—apart from Swedish—English is the main question, and that is where the 'secretary shoe' pinches. My demands are immense: first-class English, French,

German and Swedish, stenography, and the ability to use a Remington typewriter, etc., etc. But I am not totally impossible, and if I get on with the person in question, I can allow many of my demands to collapse like a house of cards. Some time ago, I engaged a Mr Sohlman, but after a week it became obvious that he was better suited to chemistry. He is now employed at my laboratory in Bofors, and is one of the few people I am really fond of. That is because even though I myself am a sort of foolish grumbler, I can understand and appreciate the value of others. *Avis à la lectrice.*

Nevertheless, not even Miss Ahlström came up to scratch, and she only worked for a time with Nobel.

After completing the rearrangement of the library and the files, I was transferred to San Remo, to work as a chemist at Nobel's private laboratory there, together with an Englishman named George Beckett, slightly older than myself, who after studies at a German university had been employed with a large-scale English chemical industry under Sir Alfred Mond (later Lord Melchett). A young Frenchman, Alphonse Tournaud, worked as laboratory assistant and mechanic. He had come from the laboratory at Sévran when Nobel's former assistant, Fehrenbach, had refused to leave France when the laboratory was moved to San Remo.

The laboratory itself, a long one-storeyed building which stood in a large park and garden adjoining Nobel's villa, consisted of three rooms: a big machine room with a gas engine and electric generators for different types of voltage, electric current for lighting and numerous electrochemical experiments; an equally large room for purely chemical tests and other experiments; and a smaller one containing a library, weighing machines and various instruments, as well as rifles for shooting practice. The firing took place towards the open sea, from a steel jetty built out from the beach; a chronograph for measuring velocity was placed inside the laboratory.

A few weeks after my arrival in San Remo, Nobel came to stay in his villa, and spent several hours each day in the laboratory, where he followed the work in hand and gave instructions for its completion.

What interested him chiefly during the following winter was experimenting with different types of smokeless gun-powder, and also a new kind of safety fuse for non-military blasting. This was Beckett's job, while it fell to me to experiment with synthetic rubber. Nobel's idea, for which he sought a provisional patent, was to produce substitutes both for rubber and leather, based on industrial nitrocellulose which was converted into the form of an elastic or more rigidly gelatinized substance by means of treatment with suitable non-volatile gelatinizing agents. It was up to me to search out and test hithero unknown gelatinizing agents, usually in connection with the synthetic production of different chemical combinations—an absorbing task for a young chemist. Results were very promising, and Nobel believed he was on the point of establishing an important industry, for which he had great expectations. Sad to say, these were not realized during his lifetime: despite its promising appearance the substitute for rubber did not come up to standard, and the factory which Nobel planned to set up for the production of artificial leather at Gullspång in the west of Sweden never materialized. Nevertheless, the results achieved later proved useful in other ways, as for instance in the production of a certain kind of such leather, the so-called pegamoid, and in the ingredients in modern cellulose lacquers.*

Apart from his work in the laboratory, two important matters took up much of Nobel's time during the winter of 1893–4, namely written negotiations in connection with the purchase of Bofors, and also the cordite process. The paper-work resulting from all this irritated him greatly. Other

* During the Second World War, the production of substitute materials for leather and linoleum profited to some degree from the original findings in the laboratory at San Remo.

involvements, of which I myself knew nothing at the time, included letters to lawyers in Vienna. He began to suffer from attacks of migraine, when he had to sit and write with his head swathed in wet towels. When the pain and his nervous depression became too acute, he would stop working on immediate problems and, by way of a change, take up his own literary efforts—among them *Nemesis*, *Beatrice Cenci* and *The Patent Bacillus*—a parody on the cordite affair.

One of Nobel's favourite recreations was to go for drives in his carriage. He owned a handsome turn-out and, both in Paris and San Remo (later on at Björkborn) kept a stable of thoroughbreds. When he and I went out driving alone, he would often begin speaking Swedish; it was probably a relaxation for him occasionally to use his mother tongue.

I can recall a couple of small anecdotes which characterize Nobel's generous way of treating a young subordinate.

Soon after his arrival at San Remo in the beginning of December 1893, he asked me to let him know what it had cost me to move back to Europe from the States. I told him I had not felt justified in asking for any recompense for this, since the Hamburg–America Line had given me a free trip to Europe as a journalist—during the Exhibition in Chicago I had done some work as correspondent to a Swedish evening paper. He looked surprised—I thought almost annoyed—and I wondered whether he thought my temporary job as a newspaperman in some way compromising. His own experience of the press had not always been happy.

However, after a day or two he presented me with a cheque for three hundred pounds, which he told me it had cost my colleague Beckett to move with his family and furniture from England to San Remo. Nobel could see no reason why my move should cost him less. I was overwhelmed by the size of the cheque, and did not know how to thank him. Another gesture of this kind was typical of his thoughtfulness for others. When the experiments with 'artificial rubber' and leather began to show promise, Nobel came into the laboratory one day and presented Beckett and me with twenty-five shares—at £10 each—in the Nobel Dynamite Trust Fund, in recognition of our work. Another incident, however, which made me feel a bit of a fool, occurred a little later. I had been given the job of trying to produce an organic compound, phthalic acid, by means of the oxidation of molten naphthalene with warm air in the presence of boiling concentrated sodium hydroxide (caustic soda). When I started the apparatus for this experiment, Nobel stood beside me to watch the result. Through my clumsiness in assembling the apparatus, warm caustic solution ran out through a rubber tube and soaked one of Nobel's trouser legs. We realized the mishap at once, though not in time to save the situation by splashing the material with water—and Nobel, obviously shaken, left the laboratory in a hurry, while I stood there both ashamed and horrified, fully expecting the sack. We waited in vain for him to turn up again for the next few days. Finally, the staff at Villa Nobel told us that he had packed a suitcase and gonc to Monte Carlo and Nice for a week, presumably to relax and forget the incident.

It is easy to imagine the feelings of a young man in my position in response to this attitude on the part of a respected employer—feelings of appreciation and affection which only increased with the years.

Early in 1894 Consul General J. W. Smitt came to San Remo with his eldest son, August—my second cousin—who was to stay on for a couple of months after an attack of pneumonia. Nobel and Smitt knew each other well and had corresponded from time to time about business matters when Smitt was—in fact, if not actually in name—managing director in the Nitroglycerine Company, of which he was also chairman. Nobel respected Smitt as a practical businessman, but their personal relations were not particularly close, owing to differences in character and outlook. Nobel thought Smitt narrow-minded in economic questions, and altogether too inquisitive about the improvements and new inventions which he, Nobel, was carrying out, and which Smitt seemed to think might affect the interests of the company. I found out later, too, that Nobel was afraid that Smitt might try to take advantage of my position by questioning me about the current work in the laboratory.

In the middle of January 1894 Nobel told me he was going away for a few weeks, first of all to London for the conclusion of the cordite process before the House of Lords, and then to do some business in Sweden. He had just signed an agreement with the owner of Bofors for the purchase of the controlling interest in the Bofors/Gullspång Company.

From what has transpired since, it appears that Nobel's annexation of Bofors had been preceded by lengthy written negotiations. His aim was to set up a factory in Sweden for the manufacture of cannon and other war materials where he could carry out experiments on a large scale, beyond the limited possibilities offered by the laboratory in San Remo and without the interference from authorities to which he had been subjected in France. There is no doubt, too, that a renewed affection for his homeland and his wish to establish a permanent *pied-à-terre* there played a part in his decision. To begin with, the negotiations concerned the purchase of Fingspångs Bruk ironworks which had been offered for sale to him by the then owner and estate manager, Ekman.

He actually went to see Finspång, but found the contrast between the stately seventeenth-century castle and the old-fashioned and out-of-date workshops discouraging. Through an estate agent, Count Wirsén, he became interested in Bofors, Kjellberg and Sons, the shares in which were held by a well-known firm in Gothenburg. At the beginning of January, after careful investigation into the possibilities there and a considerable reduction of the original price of the shares, Nobel paid a million crowns in cash for all the ordinary shares, nominally at one and a half million, plus 700,000 crowns for 5 per cent preference shares in the company. A further 300,000 crowns in preference shares were kept for the time being by the Gothenburg firm, and taken over later by Nobel.

In view of Nobel's proposed absence from San Remo at

this juncture, I applied for a short holiday to visit Sweden and Norway. One reason for this was to see my mother, and another to announce my engagement to a Norwegian girl, Ragnhild Ström, whom I had met the previous autumn. I wanted now to meet my future in-laws.

The holiday was granted, on condition that I should contact Nobel at Bofors. I arrived there on 18 February, a day after Nobel himself, and found that rooms had been reserved for him in one of the detached wings beside the manor house. This building appealed to him much more than the castle at Finspång, besides which the workshops at Bofors, even though fairly modest, were in a much better condition.

A meeting of the shareholders in the company was held at Bofors on 19 February, at which Nobel was appointed chairman of the board.

A small episode during one of those days is evidence of how easily Nobel could become suspicious, often for very little reason. One morning while we were discussing artificial silk he told me he had heard that a Swedish newspaper had recently published a description of the procedure which seemed to him to tally with certain experiments we had carried out in San Remo. I myself had been doing this work, and Nobel asked me with some irritation whether I had spoken of it to Smitt—he suspected that this was how the matter had come up. I said no, of course not, and assured him that I fully understood the necessity of never discussing his projects. But it was not until I managed to get hold of the paper in question, and could show him that the article only covered the early manufacture of these synthetic fibres by the Frenchman Chardonnet, that he calmed down. After that, he never mistrusted me again, and reacted strongly to any suggestion that he had done so. An example of this occurred during a meeting at Bofors to discuss the construction of a water-gas generator to be used for forging on the estate. One member of the group was an engineer named Carl Delwick, who had been my employer in Chicago for a year or so, and who was a specialist in the construction of water–gas generators. Delwick outlined his plans and, after listening for a while, Nobel remarked: 'I'd like to know what Sohlman thinks about this.' When someone suggested tactfully that it might be awkward for me if I had to criticize or contradict my former manager, Nobel flared up: 'Let me tell you that if the devil himself were to say that Sohlman was not completely straightforward with me, I should call him a liar!' As often happened when he thought he had gone too far, he regretted his outburst at once and continued: 'Come on, boys, let's talk about something else.' Then followed one of his frequent paradoxical sallies—scarcely worth quoting here, but very typical of him—paradoxes in regard to inventions and improvements which could make an unprepared listener wonder whether he had gone off his head, but which Nobel himself chuckled over as madcap notions—intended, as he said, '*pour épater le bourgeois*'.

After leaving Bofors I went on to Stockholm, where Nobel had instructed me to work for a few days at Alarik Liedbeck's laboratory on the examination of samples of urine taken from patients with fever, kidney trouble and syphilis—samples which were provided by one of Stockholm's big hospitals. These investigations were too superficial to yield any definite result, and I only mention them in passing as an instance of Nobel's many-sided interests. I returned to San Remo in March, to continue experiments with rubber, synthetic leather and artificial silk.

Close to Villa Nobel was another larger villa owned by an Italian named Rossi, who was eager to sell it to Nobel. To this end he made incessant complaints of the risk to Nobel's neighbours caused by his laboratory and the dangerous work that was supposed to be going on there. Tired of his perpetual nagging, Nobel ended by buying the house, which contained twenty rooms and stood in a large park on the edge of the Mediterranean. He himself had no need, or indeed any practical use for the property and, as a good businessman, felt some embarrassment at having given in to pressure. A few days later he told me he had thought of a use for the villa: 'It will be an excellent place to undress in when we take our daily dip in the sea below; that will save us from being crowded out by Italians in the public bathing huts.' (In May and June we usually bathed on the San Remo plage.)

The elegant villa was in fact never put to much use as a bathing hut, and remained empty and unfurnished until Nobel's death. The magnificent suite of furniture which he ordered in 1896 was not even ready for delivery when he died. There is reason to believe that he was thinking of offering King Oscar the villa as a residence during his spring visits to the Riviera.

Work at the laboratory in San Remo went on at top speed until the beginning of July, when Nobel decided to go to Paris and later to Sweden. I was sent to Stockholm to continue experiments with artificial leather at the Nitroglycerine Company's factory at Vinterviken, south-west of Stockholm, and then went to Bofors, where Nobel himself arrived in August. While he was there it was decided to set up a new laboratory for experimental work on a larger scale than was possible in San Remo. Alarik Liedbeck, whose laboratory in Stockholm included an engineering department, was asked to supply the plans for buildings, machinery and other apparatus, in collaboration with me.

The new laboratory was to be sited at Björkborn, a property on the Bofors estate not far from the old manor house which had been reserved for Nobel as chairman of the board during his visits to the ironworks. Construction was begun immediately, so that the plant could be used successively from the spring of 1895 onwards.

In view of the inevitable break which this would cause in our work, I asked Nobel's permission in August 1894 to go to Norway to be married. He sent me a delightful telegram of congratulation, in which he added: 'I may possibly have forgotten to mention that from 1 July your salary has been increased to 10,000 lire a year.' (My salary at that time was 5,500 lire.)

A letter written the same day ran as follows:

Paris 1894

Dear Mr Sohlman,

The telegram which precedes this letter contains every good wish for happiness to you and your bride. I added a few lines on business matters since married life involves new responsibilities, and such things should be clear from the start. I hope that in future you will always find me ready to appreciate the merits of others.

Your affectionate friend
Alfred Nobel

During the autumn and winter of 1894, when the laboratory was finally completed, I was engaged in closed vessel tests for the determination of the gas pressures generated by firing propellant powders at various loading densities. (The results proved useful in connection with the application of the so-called Bofors method for internal ballistic calculations.) We also experimented with the production of artificial gems, through the melting of clay in a vessel lined with platinum, and using explosive heat in accordance with a somewhat over-ambitious project of Nobel's—at least in view of the means at his disposal. The success of these experiments was negligible and, during the following year, continued efforts led to yet another accident in which an assistant engineer called von Feilitzen was badly burned at the Björkborn laboratory through the spontaneous ignition of a mixture of aluminium and aluminium perchlorate.

In January 1895 I was summoned to meet Nobel in Berlin for discussion. He asked me whether I would prefer to go back to San Remo or to stay on at the laboratory in Björkborn as supervisor. I answered that since he was good enough to let me choose, I would prefer Björkborn and Bofors, where working conditions were more attractive. Thus I became 'a Boforsian'.

It would take too long to elaborate on the many experiments based on Nobel's ideas which were carried out at Bofors and Björkborn in the following years; the most important of these have already been presented in the Nobel biography by Schück. An abbreviated list here, however, shows some of the varying fields covered by his experiments during that period: new forms of gunpowder, among others so-called progressive powder, blasting charges and fuses for projectiles; sealing girdles for projectiles; topographical mapping through photography by means of a camera shot up by rocket and supported by parachute; light metal alloys; electrolytic production of potassium and sodium.

To technicians in general and owners of rural iron works at that time some of these ideas seemed sheer fantasy, and Nobel's scientific judgment was often impugned. In this connection I feel justified in quoting from Schück's biography:

'It must not be forgotten that most of what contemporary scientists regarded with scepticism was nevertheless carried into practice by Nobel, and came to be of the greatest practical importance; and we must also remember that some of his other ideas came to be applied in quite a different direction and in an entirely different field than that which their inventor originally had in mind.'

In his creative activity, the true genius is as lavish with his ideas as Nature is fecund. In general, only a few of these ideas immediately find the right soil for growth and development. Some are merely barren seed. Others fall, perhaps, on stony ground because the time is simply not ripe for them. Yet buried thoughts can retain their life force for years, sometimes even for centuries. And when other circumstances and better conditions arise, they can begin to sprout, like seeds which the wind tosses onto fertile soil.

Nobel spent most of the two following summers at Bofors, staying in the manor house which he had asked his nephew, Hjalmar Nobel, to refurnish. He took a keen interest in the extension of the works, and was in constant touch with the estate manager, Jonas Kjellberg, in the planning and reorganizing. During 1894–6 when he was both owner of Bofors and chairman of the board, he took over the directing, financing and running of the entire company. There is no doubt that the basic conditions for the remarkable development of Bofors during the past fifty years can be attributed to Nobel, and that his leadership constituted a turning point in its history.

The workshops were supplemented and modernized in both buildings and mechanical equipment; the manufacturing capacity increased through the acquisition of new constructions such as heavy cannon; gunpowder production and other chemical industries were set up; and the financial status of the works was completely restored, due to the investment of fresh capital by Nobel, who subscribed two and a half million Swedish crowns in shares.

Nevertheless, Nobel's enthusiasm for Sweden was not limited to Bofors. He showed a personal interest in the experiments of a number of Swedish inventors, and helped many of them financially.

In 1891 he encountered an inventor in the military field, Captain W. Unge, who had constructed and patented a telemeter—an instrument for measuring distance—which had been introduced by the Swedish Army Material Administration in 1886 for use by the artillery. Unge had designed this instrument in a small workshop where he also made many other experiments. To achieve a sounder financial basis, he tried in 1891 to form a company with a share capital of 40,000 Swedish crowns, and managed to interest several prominent people in the scheme, among them King Oscar of Sweden. The Nitroglycerine Company also subscribed, and he finally approached Nobel, who made up a deficit of 7,000 crowns for the new company, which called itself Mars, Ltd.

This was also the beginning of a collaboration between Nobel and Unge on another military invention, namely the improvement of rockets, in particular war missiles. In July 1892 Nobel had taken out a provisional English patent for the construction of such missiles, and in September of the same year, when Unge visited him in San Remo, the two men signed an agreement to co-operate on this and other inventions in the same field.

The practical side, the making of these products, was to be Unge's responsibility, while Nobel was to assist in exploiting them; he would also shoulder the costs, both for the experimental work and the patenting. From the eventual profits, Nobel would retain two thirds and Unge one third.

Subject to the agreement, the perfecting of these inventions continued until Nobel's death and, with the help of the executors, for some time afterwards. The idea of the guided missiles was later bought up by Krupp and further adapted in Germany. One may surmise that the Nobel–Unge rockets, air torpedoes as they came to be known, which were of such paramount interest to Nobel—not only from the military viewpoint, but also as safety equipment for ships—were forerunners of the dreaded V1 and V2 weapons in World War Two.

An engineer named Strehlenert who had invented an apparatus for pressure nozzles used in extruding rayon fibres was given generous support by Nobel for his method. In this connection Nobel himself patented a method for producing pressure nozzles with extremely small holes which later had some bearing on the development of artificial silk manufacture. The Swedish costs incurred by the Ljungström brothers for the promotion of various inventions—chief among them the Sveavelocipeden (Svea bicycle), the first bicycle with variable gears—were underwritten by Nobel. He also helped to found an English enterprise, the New Cycle Co., for large-scale exploitation of the Svea bicycle, buying shares in this company for a total of 75,000 crowns.

It may also be mentioned that S. A. Andrée's balloon expedition to the Pole was made possible by the financial help he received from the inventor. After the failure of the first attempt in 1896, Andrée came to Bofors to meet Nobel. My wife and I, who had known him earlier, were invited to lunch with them, and I well remember how warmly Nobel congratulated Andrée for not attempting the flight when conditions were unfavourable, at the same time promising his support for a new expedition.

In the spring of 1895 Nobel met a civil engineer named Rudolf Lilljequist who made such a singular impression on him during six months or so of negotiation that he appointed him as an executor of his will. Lilljequist himself has given a detailed account of the circumstances which led to this friendship, and Nobel's interest in the business Lilljequist intended to establish in Sweden. He writes:

> After many years abroad, in France and England, I decided at the age of forty to come back to Sweden, to obtain a firmer foothold there than I could hope for in a foreign country.
>
> It is no easy matter for a middle-aged engineer, whose practical experience is based solely on French and English conditions, to find the fleshpots of Egypt among his own people.
>
> It was thus essential for me to take up some form of independent work with reasonably good prospects in the country. A good idea seemed to be the exploitation of that country's water power.
>
> In the spring of 1894 I had read in a magazine called *Engineering* of a new method developed by Castner to decompose common salt electrolytically by employing a mercury cathode, and I found the subject interesting. A first essential was, of course, to test and work out the details so that it could be adapted to Swedish conditions.

Experiments were carried out at the Institute of Technology in Stockholm, under the supervision of an expert in electrochemistry, Dr Cassel, who confirmed that the method seemed very promising. The choice of a site for the proposed installation fell on Bengtsfors, in north-western Sweden, where a suitable waterfall existed. Lilljequist added:

'The main difficulty was how to raise the necessary capital for the establishment of the enterprise.

'A friend advised me to write to Alfred Nobel in Paris. "He is the man for your plan," he urged. I put the matter to Nobel, and received the following reply':

Dear Mr Lilljequist,

Yours of the 28/2.

It has been in my mind to make use of the Gullspång waterfall in roughly the way your memo suggests. It seems to me, however, that our experience regarding the manufacture of bleaching powder, caustic potash and caustic soda is still insufficient and also unsatisfactory, and I therefore hesitate to recommend any larger installation at present, although in these days things move fast. The difficulties have been pretty well overcome regarding potassium chlorate. They were considerable in the beginning, and the Schweitzer Company spent large sums just buying platinum. Now this production seems to be paying off.

Awaiting further information and proposals,

Yours faithfully, etc.

After receiving the information asked for, Nobel announced that he would come to Sweden in the summer for further discussion. He met Lilljequist in Stockholm in May 1895. As a result of their talks, which included a thorough analysis of electrolytic problems, Nobel agreed to invest 100,000 Swedish crowns in the new company, on condition that the remainder of the necessary cash deposit, estimated at 300,000 crowns, could be found elsewhere. As compensation for the right to the method, a total of 50,000 crowns was to be deposited in the form of ordinary shares.

A little episode which occurred during Nobel's first meeting with Lilljequist is characteristic of Nobel and his dislike of being disturbed while engaged in some topic which interested him. As they were discussing electrolytic problems, word came that the famous banker, Knut Wallenberg, was in the waiting room; having heard that Nobel was in Stockholm, he had come to see him about some investments. Nobel sent a message that he had no time for visitors, and the investments could wait. Wallenberg, who was a great friend of Nobel's and knew his working habits, was not offended —but Lilljequist suggested that he could come back another day . . .

As far as I know, Lilljequist and Nobel only met once or twice, but they kept in touch and their negotiations continued by letter. In June 1895 Nobel confirmed his belief in the possibility of an important electrochemical industry in Sweden. He himself was worried about the production of caustic soda according to the method described by Lilljequist, and suggested that unless the latter found the experiments entirely satisfactory they should be continued at Bofors. In any case, Nobel intended to set up an electrochemical laboratory of his own at Bofors for the practical testing of promising new electrolytic methods.

The letter is proof of Nobel's caution in regard to the adoption of new manufactures on a large scale, and also of his determination to convince himself by means of his own work. In a lengthy answer to Nobel, written in June, Lilljequist pointed out, after giving a detailed account of the experiments already made, that in his view they seemed entirely satisfactory, and it was now time to go ahead with construction on a larger scale, a decision which Nobel had earlier expressed himself willing to support. The condition laid down by Nobel for his investment of 100,000 crowns, namely that the outstanding 300,000 in cash should come from other interested parties, was met by Lilljequist and his relatives in Switzerland who bought shares for 70,000 and placed the remaining 230,000 in Denmark and Sweden. The subscription list was signed by Nobel on 25 July 1895, and the new company approved by Royal Ordinance on 9 August the same year.

The complete confidence which Nobel placed in Rudolf Lilljequist was further emphasized when, on the retirement of Jonas Kjellberg from the works management at Bofors in 1896, Nobel offered Lilljequist the post on extremely favourable terms. The latter did not, however, feel able to accept this flattering offer on account of his engagement in the Bengtsfors enterprise. He heard nothing much more from Nobel and, like myself, did not know about his appointment as an executor until after Nobel's death.

One may perhaps wonder how Nobel, with his very critical disposition, could be so immediately impressed by Lilljequist. His interest in Lilljequist's plans to establish an electrochemical industry in Sweden was in line with his own idea of developing the Gullspång waterfall. On meeting Lilljequist Nobel must not only have formed a high opinion of the man's capability and ambition, but must also have taken a strong liking to him.

Lilljequist and Nobel had much in common, both being expatriate Swedes with a marked international profile. Lilljequist gave the impression of an upper-class Englishman, rather than a Swede. He spoke and wrote fluent English, and Nobel complimented him on his command of the language. They agreed on many subjects and both criticized the Swedish interpretation of the rights of foreigners to own shares and become members of the board in certain Swedish companies. They also shared a mutual distrust of lawyers. In dealing with a troublesome water-rights problem in Lilljequist's enterprise, Nobel wrote:

'Lawyers have to make a living, and can only do so by inducing people to believe that a straight line is crooked. This accounts for their penchant for politics, where they can usually find everything crooked enough to delight their hearts . . .'

The moods of extreme depression evident in Nobel's letters during the latter half of the 1880s and, in particular, the years 1890–93, compared with his general attitude towards the future and the people around him in the last three years of his life, would seem to show that in spite of increasing heart trouble he was now much happier and more balanced than earlier. This may have been due to closer contact with his homeland, and also to his discovery of fresh and stimulating goals. He had been able to realize his early dream of creating a home—or rather two homes—for himself in beautiful surroundings, and had also made new friends. His previous feelings of isolation may have decreased. Meeting him daily during those last years, I saw no sign of the nervous depression which he used to call 'visits from the spirits of Nifelheim', which had been particularly frequent in the winter of 1893–4 in San Remo.

In the autumn of 1895 Nobel spent about two months in Paris, working on the provisions of the will which became of such fundamental importance in the creation of the Nobel Foundation and its various activities. This will is dated 27 November 1895 and was clearly composed and written by Nobel in his Paris home without any assistance. He signed it in early December at the Swedish Club in Paris, in the presence of four Swedish witnesses.

With this will, Nobel declared all his previous testamentary dispositions rescinded. The only disposition which remained from a will dated 14 March 1893 was a clause appointing the Academy of Sciences in Stockholm as chief beneficiary, and distributor of prizes 'for the purpose of forming a fund, the interest on which shall be distributed by the Academy each year for the most important and original discoveries or intellectual achievements in knowledge and progress, excluding physiology and medicine' (where awards were to be distributed by the Caroline Institute).

I shall later compare the two testamentary documents in the light of further experience, and look at the motives which may have prompted Nobel's readoption of the first of them. The basic idea is the same in both cases, namely that the bulk of his fortune should be allocated to the support of scientific research and pioneering work 'in the field of knowledge and progress'—including the peace movement—and that responsibility for carrying out the testator's wishes was to be entrusted to Swedish institutions (through the last will) in co-operation with a committee appointed by the Norwegian Parliament (Storting).

It is probably safe to assume that Nobel's realization that his indefatigable work had at last found a goal—on a higher plane than his successes at Bofors—made him more hopeful in later years.

There had been times when it seemed that the lack of such a goal was a torment to him. His work had gradually become a habit, without ultimate purpose.

His eagerness to plan the immediate future and to reorganize his life during the very last year may seem strange. His aim was to spend most of the summer at Björkborn, and with this in mind he brought a French chef (Swedish cooking did not appeal to him) and some of his staff from San Remo. The refurnishing of the manor house has been mentioned already; in addition he installed a stable of three Russian horses, a carriage and a groom. In the autumn of 1895 King Oscar intimated that he would like to see Bofors and meet Nobel. The royal visit took place in September, when the king did a round of the workshops and lunched with Nobel at Björkborn. (It may have been this visit and the possibility of future meetings in San Remo which lay behind the renovation of Villa Rossi.)

As time went on, Nobel became anxious to re-establish his links with Sweden, even though contact with his homeland had never really been broken. Apart from writing to his mother and brothers he had corresponded on business matters with a great many Swedes, chief among them his old friend and colleague Alarik Liedbeck.

In 1880 Nobel donated 50,000 crowns from his maternal inheritance to the Caroline Institute 'to form a Caroline Andrietta Nobel Fund for experimental research in all branches of medical science and for the furtherance of such research, both for instructional and general purposes'. In this connection he announced that he would like to contact a Swedish physiologist regarding certain investigations. A laboratory chemist, J. E. Johansson, who was then in Paris and heard of this wish, immediately contacted Nobel; he later worked for five months at the laboratory in Sévran, experimenting with blood transfusion, a subject which greatly interested Nobel. During talks with Johansson, he often mentioned that he would like to found an institute of his own for medical research. This, and similar comments which Nobel made to me directly, and which I interpreted as his intention to found a technical-scientific institute for experiments in chemistry and other fields, has given added incentive to my work on the development of the Nobel Foundation.

At San Remo, Nobel also renovated and improved his home during his last year. He wanted to spend the winters there and to some degree co-ordinate the San Remo and Björkborn laboratories. Soon after his death, I was surprised to find a small house in the grounds which had evidently been built for me and my wife as winter headquarters, although he had never told me about it.

Further evidence of how little attention Nobel paid to his heart condition at that time—he even joked about it in letters—was that he changed his stable, and bought three handsome carriage horses, one of them a fine bay. He paid 13,650 French francs for them.

Nobel spent much of the summer of 1895 at Björkborn, with a break in June and August for trips to London and Paris, and for a short holiday in Karlsbad to take the waters. His programme was much the same in the following year. The summer months at Björkborn were devoted to the planning and extension of Bofors, the purchase of smaller ironworks in the neighbourhood, and work in the laboratory. It was in 1895 that he suggested that we should call each other by our Christian names—a gesture which I, as a young man of twenty-five, at first found highly embarrassing. In Sweden at that time it was not usual to be on such familiar terms with a superior—and I knew that generals and colonels who had made the same suggestion to officers of a lower rank were still addressed as before by their subordinates. For me, Nobel was at least a general! I therefore started off by calling him

'Doctor', which seemed to thoroughly amuse him, and he told me to stop it. (In 1892 the degree of Doctor of Philosophy had been conferred on him by Uppsala University, a distinction which pleased him greatly.) He did not even like to be addressed in letters as 'Dear Colleague', so I began writing 'To Mr Alfred Nobel', signing myself 'your affectionate friend R . . .'. If I happened to write 'your sincerely affectionate friend', he pointed out that the word 'sincerely' was superfluous!

Nobel and my wife became great friends, and he asked her to translate his own work *Nemesis* into Norwegian. Now and again, she would be invited, as the only woman, to dine with the governing body at Bofors.

While Nobel was away from Bofors I sent him reports of the on-going work there, receiving answers with instructions by letter and telegram. On 6 April 1896 he wrote:

> My health is unfortunately giving me a lot of trouble this year, and much of what should be done has been neglected. Cable me if you think you can get away to San Remo for, say, a fortnight. Your wife's ticket will of course also be taken care of. If so, I will invite Oskar Ljungström, as I have many important matters which I am anxious to discuss. They concern gun forging, artificial rubber and much else.
>
> Your affectionate friend, A. Nobel

I answered that the work in hand would not prevent me from going away for a week or two, and received the following telegram: 'Come. It will be refreshing and useful, but no explosives are to be manufactured in your absence.' My wife and I arrived at San Remo at the end of April and were warmly welcomed by Nobel. We stayed for four weeks, during which I worked in the laboratory on various projects, on the specification of patents and so on.

In due course, however, I began to feel that the experiments with dynamite at Bofors were more important than the work I was doing in San Remo, and asked Nobel's permission to go home. He finally agreed, although I realized later that he would have preferred us to wait and travel back with him at the end of May. Now, long afterwards, I have understood that he would have appreciated a more easygoing and less 'respectful' attitude on my part than I felt able to show him. He gave me so many proofs of his friendship and confidence, which I did not at the time dare to interpret except as evidence of his natural good will. In a letter dated February 1895—speaking of the acceptance of suggestions from outsiders—he ended: 'There are two things which I never borrow—money and projects. But if anyone as sound as Mr Sohlman is willing to lend me a little friendship, I shall accept it with gratitude.' On another occasion, he said to me: 'You know, Ragnar, I almost think of you as a young relation.'

I quote these words, not in any way as a compliment, but only to stress the longing—indeed the need—Nobel obviously felt to have someone close at hand who was fond of him, and on whom he could depend.

After the summer together at Björkborn, when Nobel was back in Paris, he wrote to me on 25 October 1896:

'My heart trouble will keep me here in Paris for another few days at least, until my doctors are in complete agreement about my immediate treatment. Isn't it the irony of fate that I have been prescribed N/G 1, to be taken internally! They call it Trinitrin, so as not to scare the chemist and the public.' He added a postscript: 'Tell your charming wife that I have fallen in love, not with Björnson, but with his writing.'

His last letter to me was written on 7 December 1896. It was still lying on the desk when I arrived in San Remo after his death, and ended with the words:

> Alas, my health is so poor again that I can only scribble these words with difficulty. But I shall come back as soon as possible to the subjects which interest us both,
>
> Affectionately, Alfred Nobel

His handwriting was clear and legible as usual, and gave no suggestion of imminent collapse. But soon after ending the letter, he appeared to have had a warning of the cerebral haemorrhage which was to end his life, at two o'clock in the morning of 10 December 1896.

Nobel's last hours were shadowed by sadness. The fears of solitude he had often expressed in letters were confirmed, and his only companions at the end were servants; there was no one close to him at hand 'to close his eyes and whisper a gentle and consoling word'. The servants helped him from his study to the bedroom on the second floor of the villa, and his Italian doctor was summoned, though only to confirm the gravity of his condition and give orders for complete rest in bed. These, however, were difficult to carry out, since the sick man often became violently agitated, and had to be forcibly prevented from getting up. His oldest retainer, Auguste, reported later that he was unable to speak, and could only mumble some words in his mother tongue, which none of them understood. All they could make out was that cables must be sent to his family and to me.

The news of his sudden illness reached me on the morning of 8 December and after contacting his nephews Emanuel and Hjalmar I made arrangements to go at once to San Remo. Emanuel and Hjalmar decided to go too, and we met *en route*, arriving during the night of 10 December too late to see him alive—the cable announcing his death had reached us on the way.

Alfred Nobel died, as he had lived, alone.

On 11 December 1896, only three people closely connected to Nobel were present in San Remo—Emanuel, Hjalmar and myself. Nothing was known then about the dead man's will, or its provisions. Emanuel Nobel, his oldest and closest relative, took charge of the immediate arrangements for the funeral. After a simple ceremony in the villa, it was decided that the coffin should be sent to Stockholm. The Swedish pastor in Paris at that time, Nathan Söderblom (who later became Archbishop of Sweden) had been a friend of Nobel's, and was now asked to officiate at the service in Villa Nobel. He agreed and, on the journey from Paris to San Remo, left the train at Nice, to walk from there to Menton along La Grande Corniche—the mountain road high above the coast-line, with its magnificent view across the glimmering Mediterranean—no doubt contemplating the speech which he made on 17 December beside Alfred Nobel's bier, and from which the following excerpts have been taken:

> Before Alfred Nobel begins his last journey to the nordic country which, although very much a citizen of the world, he always looked on as his own, we, his friends and those close to him gather here, between the mountains and the sun-tipped waves of the Mediterranean, in the earthly paradise he loved so deeply that he chose it as his home.
>
> It is this home which has been the springboard of his creative genius and his untiring work—for a restless spirit like his had little time for the peaceful relaxation sought by many who come here. It is here that the light of his earthly life has suddenly been extinguished. Today, this home has been transformed into a house of mourning. We stand at this moment before the stern face of eternity, or perhaps I may say in the sight of the Almighty, Master of life and death.
>
> Many words will wing their way to Alfred Nobel's bier. Words that recall the dead man's immense intellectual power, his remarkable achievements, and his conquests of Nature's hidden forces in the service of humanity; the honour he has brought, throughout the world, to our distant country; the hard-won fortune, so much of which he spent on the furtherance of culture and the relief of suffering and need; and his shrewd and passionate aptitude for research, which aimed at exploring the farthest reaches of human endeavour.

And yet, at this moment, he would not have wished our thoughts to linger on any of these things.

By a strange coincidence, some weeks before his death, he allowed me to read some lines from a manuscript he was working on. Soon after the sad news reached me from San Remo, I came upon a few words on one of its pages, which he scarcely imagined would so soon apply to himself—but which we have every right to quote here, since they give us an idea of his own reflections on life and death.

They read: 'In stillness, you stand before the altar of death. Life on earth and life beyond the grave are eternal mysteries; but our own dying embers must give us pause, and silence every voice except that of religion. It is eternity speaking.'

Thus, beside this bier, the clamour of fame and extravagant speeches of praise are hushed. Death knows no difference between the millionaire and the penniless, between the genius and the simple-minded. At the end of the day, we are all equals. In death, as in religion, it is the soul alone that counts.

It was a natural corollary of the loneliness and suffering that were his lot that in the public estimation he should have figured so much as a rich and remarkable man, and so little as a human being. Let us not perpetuate this error now that he is dead; for to the life beyond the grave we can take neither our possessions nor our achievements, and we must leave behind our earthly happiness too. In such happiness our dead brother may well seem to us, despite all his possessions and the affection of his associates, to have been poor indeed. It was his choice, or his fate, to live alone, and he died alone, without a hearth to cheer him or the hand of a son or a wife to smooth his brow. And yet his was not a nature to be hardened by money, or to be embittered by loneliness; to the end of his life, he was warmhearted and kind. In the life beyond, all that matters is to have lived nobly.

At Christmastide, we northerners turn towards home. Even the smallest and poorest corner which can be called a home holds a special glow and attraction. Everyone with true feeling, or who has any family sentiment, longs to share with those close to them the warm light of the Christmas candles that shine through the darkness of our nordic winter.

Now Alfred Nobel has made his final journey through the Valley of the Shadow. On the other side a home awaits him where, through God's great mercy, everlasting Christmas light shines over life and death. Nobel called this an eternal mystery, although he himself always sought an answer to this mystery through work and generous dealings.

As he himself has said, every voice except that of religion is silenced before the altar of death. The true voice of religion is finally His, who said: 'I am the Way, the Truth and the Life'. Alfred Nobel heard these words above the tumult of the world around him. I know that he listened to them with humility and respect.

After the short farewell in the villa, the flower-strewn coffin was carried in procession to the station for transport to Stockholm. The funeral took place in the city's Great Church (Storkyrkan) on the afternoon of 29 December 1896. Flowers and exotic plants gave colour to the thirteenth-century building, and the main aisle had been transformed into a pathway of palm trees, intertwined with ivy.

Many outstanding Swedes and representatives of Nobel's enterprises abroad were present at the ceremony, and magnificent wreaths surrounded the coffin, which was later taken to the crematorium in the north of the city, preceded on its way by mounted torch-bearers.

The Latter Years of Alfred Nobel's Life

The last years of Alfred Nobel's life, in particular the period 1880–96, were marked by bouts of deep depression. His private correspondence shows how isolated he often felt, despite his restless activity and many successes in science, industry and finance. The main reasons were poor health and his disillusionment with fraudulent colleagues, but there was also another significant factor which will be discussed later.

Increasing ill health was beginning to affect his mental state. As a child he had always been delicate, and he often referred bitterly to this handicap. Replying to a request from his brother for an autobiographical note, he wrote: 'Alfred Nobel—a pitiful half-life which ought to have been extinguished by some compassionate doctor as the infant yelled its way into the world.' His health was a worry to his parents, and as early as 1854 he had to take a cure at a spa in Germany. Despite his hatred of the enforced inactivity such watering places entailed, he was often obliged to have recourse to them in later life. In his own view, holidays in the Stockholm archipelago did far more for his health than the spa, 'with its casual acquaintances whom, though agreeable at first, I abandoned with as little regret as a worn-out coat'—a remark he made at twenty-one, but which was to be typical of his later attitude to social life.

Yet during the sixties and seventies his creativity, energy and willpower were unimpaired by his physical weakness; it was in this period that he made his greatest inventions and, during continuous travels in Europe, established the companies which became the sources of his fortune. He was a man of immense perseverance, and the courage of his convictions enabled him to stand up to setbacks of many different kinds. These included a long series of accidents in his factories in various countries, resulting in great material damage and loss of life, as well as ruthless attempts by competitors to rob him both of the glory and the financial reward of his inventions. He even managed to keep up with many friends, and carried on an extensive correspondence with intellectuals and artists in Paris and elsewhere. In 1875 he completed his third important invention—blasting gelatine or rubber dynamite—the 'Nobel Detonator' being the first and kieselgur dynamite the second.The technical application of this third discovery was to keep him very busy during the next few years.

Towards the end of the seventies, his physical condition started to decline. Since he caught cold very easily he was sensitive to any change in the weather, and the Swedish climate presented a particular problem during his annual visits to his mother. He was also very dependent on his diet. French cooking suited him best, and he abhorred the heavy English and Swedish meals which he had to face on his journeys. In his private correspondence at this time he frequently complained of other forms of illness, and symptoms of ageing. In the autumn of 1878 he said that he was suffering from scurvy, and wrote that the cure prescribed—horseradish and grape juice—was doing him no good. A few years later he consulted a French doctor who diagnosed advanced scurvy and ordered more of the same diet, combined with strong salt baths, to be taken in Austria. Nobel's letters also mention severe migraine, which made it impossible for him to work and could keep him in bed for days on end. This caused him to seek escape from the city in the peace of the countryside. Returning to Paris after a short visit to Trouville in September 1878, he wrote: 'The sky here stretches like an ashen veil over the earth; high waves rage around me, but they are only a deluge of sin and folly—what a contrast to the calm beauty of Trouville! How human beings change. I used to long to live in this great city, where I could watch and share in human endeavour, but now I am hankering to get away and enjoy the delicious feeling of peace on earth which precedes eternal rest.'

As time passed, other more serious symptoms appeared, among them indications of heart disease—his elder brother Ludvig had died of a heart attack in April 1888. Nobel begins to speak of angina; these attacks became increasingly frequent and worried him greatly, especially on his travels. His feelings of loneliness and of desire for close friendship are poignantly expressed in his letters from this period. In October 1887 he wrote:

> I have been ill for nine days and obliged to stay indoors with no company except a servant, and no one to care for me. I feel I am much more seriously ill than Bouté [his doctor] believes—the pain is insistent and does not let up. My heart is as heavy as lead. At the age of 54, when one is completely alone in the world, and shown consideration by nobody except a paid servant, one's thoughts become gloomy indeed. I see in my man's face how he pities me, but of course I don't let him realize that I know.

In July the following year he wrote from Vienna to a friend who had visited him there:

> Soon after you left, I had an experience which might seem of little importance to others, but which reminded me of how sad it is to be without a friend who could whisper a consoling word, and would one day gently close one's eyes. If I could only find someone like that! It may be that I shall move back to my mother in Stockholm—there at least I have someone who is not just interested in my money . . . And now to my experience. One night at about 2 a.m. I suddenly felt so ill that I did not even have the strength to ring my bell or unbolt the door. So I had to spend some hours entirely alone, wondering if they were to be my last. No doubt an attack

of angina, a complaint I had once studied, though not in the laboratory. I have felt despondent ever since, and my heart is framed in black, like this notepaper. [Alfred was in mourning for his brother Ludvig.]

Such laments frequently recur in his letters during the next few years.

One reason for his pessimism is probably to be found in certain circumstances in his private life, to which I shall return later. But it was also undoubtedly due to disappointment over abuses of his confidence, in particular those perpetrated by his collaborator of many years standing, Paul Barbe, and by the men the latter appointed to represent the French company *La Société Française pour la fabrication de la Dynamite*, which was the centre of the 'Latin' group of dynamite factories.

Paul Barbe had an engineering degree, and had been an artillery officer during the Second Empire; after retiring from military service he joined his father's firm, *Barbe père et fils, Maitres de Forges* in Livendum (near Nancy) as a junior partner, and thus belonged to the group of leading industrialists which played such a notable part in French domestic and foreign policy during the Third Republic, *le Comité de Forge*. As early as 1868, Nobel had gone into partnership with Barbe for the exploitation of dynamite in France. Barbe was put in charge of the establishment and administration of the different companies, while Nobel supervised technical improvements and other main organizational matters, including co-operation between the companies abroad.

Nobel greatly respected Barbe's intelligence and his technical and administrative skill, but soon became aware of his unreliability whenever his own interests were concerned. In a letter to Ludvig Nobel he described Barbe as 'a smart fellow, with an excellent capacity for work, but with a conscience as elastic as rubber. A pity, since such quick-wittedness and energy seldom go together.' Nevertheless, the extensive correspondence between Barbe and Nobel, preserved in the archives of the Nobel Foundation, is clear evidence of friendship and mutual appreciation.

During the war of 1870–71 Barbe had been closely associated with Gambetta, who commissioned him to start the first dynamite factory in France to supply explosives to the French army. This friendship had largely contributed to Barbe's success in overcoming government opposition to the establishment of the private manufacture of explosives in the country. It also caused him to take part in politics himself; he joined the radical leftwing party which was supported by Gambetta as a counter-balance to the clerical and monarchical Right.

In 1885 Barbe was elected to the Chamber of Deputies as representative for the city of Paris, and in 1887 he became Minister of Agriculture in Rouvier's government. The latter fell after a year, but Barbe continued to exert great political influence as a deputy. He turned this to advantage in connection with an application for a lottery loan of 600 million francs made by the Panama Company (the enterprise founded in 1879 by de Lesseps and Baron Reinach for the construction of a canal across the isthmus of Panama). In 1886 Barbe, who was on a committee appointed by the Chamber of Deputies, strongly opposed the company's request for a loan, and it was ultimately refused by the Chamber. The chief reason for Barbe's hostile attitude was reportedly a controversy with the canal company over the delivery of dynamite, which the company decided to take from the U.S.A. instead of from France. Once this matter had been settled to Barbe's satisfaction, and he had been promised a loan of 550,000 francs by Baron Reinach (of which only 220,000 was later repaid) he initiated a campaign for the granting of the lottery loan, which was finally achieved with the support of a number of other politicians—again thanks to skilful financial persuasion. From what has later come to light in connection with the so-called Panama scandal, it is clear that the company had paid out over three million francs in bribes, of which half a million went towards party funds.

Nobel, of course, knew nothing about Barbe's manipulations, and did not in fact find out until the Panama lawsuit began, two years after Barbe's death. But repercussions from Barbe's activities as a politician and chairman of the board of directors in the Dynamite Company were to affect Nobel in several ways during 1889–92.

Thus, as a result of the conflict in the Chamber of Deputies between Barbe and the President of the Council, de Freycinet, who was also Minister of War, a violent newspaper campaign was instigated against Nobel and Barbe, who were accused of selling Nobel's discovery of ballistite to the Italian government, to the detriment of France. It was also said that Nobel, from his laboratory at Sévran-Livry near Paris, had spied on the neighbouring experimental institute run by *l'Administration des Poudres et Salpêtres*, where a form of smokeless gunpowder invented by Saraut and Vielle was being perfected for the French army. Nobel's laboratory and his shooting range for experiments with rifle and cannon powder were closed by the police, and he himself threatened with imprisonment if he continued his research on French soil. He therefore decided to leave France and move to Italy, where he bought a villa with a large area of land in San Remo.

In 1890–92 the French dynamite companies *La Société Centrale* and *La Société Générale* suffered heavy losses, partly due to dishonest speculation and partly to embezzlement by the directors, who had been close assistants to Barbe. *La Société Centrale* had speculated wildly in the purchase of glycerine, and to avoid financial collapse Nobel was obliged to arrange extensive credit for the company. Shortly afterwards it turned out that the managing director, ex-senator Le Guay, was responsible for still further illegal actions. He had also co-operated with Arton, another of Barbe's colleagues (described by the latter as his alter ego in delicate negotiations) who was later tried and condemned for forgery and other crimes.

The losses incurred were so heavy that Nobel, when

informed of them during a visit to Hamburg, believed himself ruined and, according to what one of the directors in the German Dynamite Company told me later, had actually applied for employment as a chemist in his own company.

Even though he had not taken part in the daily running of the adminstration he could, as a member of the board in the French company, be held financially responsible for the losses. After receiving further information by cable, however, he calmed down and decided to reorganize the company completely. Returning to Paris, he sacked the previous management and saved the company from disaster by means of a bond loan, a considerable amount of which he took over himself. A respected businessman was then installed as manager, who soon put the French companies back in order.

This satisfactory outcome was entirely due to Nobel, although his health was already undermined by illness and worry, and these conflicts clearly made further inroads on his initiative. On 10 October 1892 he wrote to Emanuel Nobel from Paris:

> . . . My position here is not what it used to be. I am at loggerheads with all the directors I have had to kick out; consequently, I must obtain and keep a majority of the shares, in other words 20,000 at 450–500 francs each. Even if some of my friends can help me, this is an enormous sum. If I can't guarantee it, my co-directors and I will really be in the soup, since we are having to deal with a bunch of crooked lawyers and bloodsuckers. Believe me, there is nothing more dangerous than being director of a French company . . .

In November 1892, he wrote to an English business acquaintance:

> Some days ago I was dunned for the tidy sum of 4,600,000 francs, on the ground of my alleged responsibility for the Arton–Le Guay frauds. French law is indeed curious. If insufficient control is proved, com-

> pany directors who have acted in perfectly good faith may be made responsible. The board members and their lawyers agree that no fault can be found—but when there is a court case, Wisdom herself becomes blind, and constipation or its opposite in a judge may influence his views of right and wrong . . .

Another disappointment which affected Nobel deeply at this time was the rupture of his relations with Sir Frederick Abel and Professor Dewar, in connection with 'the cordite case'. Apart from the considerable financial interests at stake, there was the crucial question of Nobel's priority as discoverer of smokeless nitroglycerine gunpowder.

In 1887–9, when Nobel was in the midst of working on his invention of ballistite, the British government appointed a special research committee, with Sir Frederick Abel and Professor Dewar as its chief members. Their task was to find some means of producing the best type of smokeless gunpowder, a vital question for most countries, and particularly the Great Powers at that time. Abel and Dewar had secretly contacted Nobel on a friendly basis, and for a whole year received information from him about his discovery and the improvements he was making. Meanwhile, they started experiments of their own with nitroglycerine gunpowder, although using a different type of nitrocellulose to that advocated by Nobel and, without the latter's knowledge, taking out a patent for the resultant smokeless gunpowder, which was named cordite. After attempts to reach an agreement failed, it was decided to settle the argument through a 'friendly suit'. The proceedings turned out to be very extensive—the printed documents alone occupy several shelves at the offices of the Foundation—and the verdict went against the plaintiff, the English Nobel Company, which was made to pay legal costs amounting to £30,000. The only consolation Nobel received was a statement by Lord Justice MacKay before the Court of Appeal, in connection with the verdict. Though agreeing with his colleagues on formal grounds, he expressed his recognition of Nobel's pioneering work as follows:

> Clearly a dwarf who succeeds in climbing onto the shoulders of a giant is able to get a better view than the giant himself . . . In this case, I can only sympathize with the original patentholder. Mr Nobel made a great discovery, and produced what in principle was a very real novelty. Two skilful chemists then got hold of his specification, studied it carefully and, in the light of their own extensive knowledge, found that they could use practically the same materials—with one exception—to achieve precisely the same result. One wishes that it had been possible to reach a decision which would not have deprived Mr Nobel of an exceedingly valuable patent.

A sharper comment on the behaviour of Nobel's opponents can scarcely be imagined—even if the most significant point in the case, i.e. the early co-operation between Nobel and Abel and Dewar, was not even mentioned. It is easy to understand Nobel's bitterness at the injustice of the verdict, and at the abuse of confidence to which he had been subjected.

For a better understanding of Nobel's depression and increasing homesickness for Sweden during the 1880s and 1890s, we must turn to the story of his long relationship with a young Austrian girl, Sofie Hess, a subject which has never before been enlarged on, out of consideration for persons still alive.

In Schück's biography, Nobel's mother is referred to as the great love of his life: 'He appears to have had no other—with the possible exception of a girl mentioned in one of the poems of his youth.' Nobel's attachment to his mother is indeed often emphasized in his letters, but Schück's statement was also clearly influenced by the said consideration.

The description of Nobel as a cold personality, incapable of warmth, is incorrect and presumably based on the fact that he never married. He was nervous and excitable, with a tendency to moodiness and violent outbursts if someone or something displeased him. 'Now the Nobel blood is boiling!' he would exclaim. From his earliest years he had an obvious need of tenderness and personal closeness which, however, was thwarted by scepticism and self-doubt.

His attitude to his mother was never affected by his moods, and their relationship was always one of mutual consideration and love. In a letter of September 1878 he describes a birthday visit to her in Stockholm:

> It gives my old mother so much pleasure to see us all assembled together . . . I myself have never known any great joy—though I have indeed experienced deep sorrow—yet I can understand the feelings of others in such matters. When life is drawing to its close, and only a few steps separate us from the grave, we rarely make new friends, and turn more confidently to the old ones. I would have loved to stay with my mother, and it would surely have delighted her—but we are too many to be put up in her house and, since exceptions cannot be made, we have had to put up at the Grand Hotel where the furniture is covered with gilt, and the food lousy . . .

A year or two later he wrote after a visit to Stockholm:

> My presence makes Mother so happy that it breaks my heart to have to leave her alone for so much of the year. When my brother Ludvig went off, she almost fell ill. At seventy-nine, one is no longer made of iron, and very little is needed to cause a physical breakdown.
>
> Since returning here, I have been overwhelmed by my correspondence, and have had to waste many hours writing letters, hours which I would far rather have spent with the old lady. She is like you . . . when she is happy she can put up with anything, otherwise the

slightest setback hurts her. Yesterday we went to a restaurant for dinner. The food didn't suit me at all and I was ill all night. Mother, on the contrary, had no trouble at all.

Nevertheless, Nobel's natural craving for sympathy and tenderness, so contrary to his seeming contempt for people in general, could not be wholly satisfied by his relationship to his mother. He must have felt the need for contact with a younger woman, and preferably an intellectual equal. Schück writes that whenever he could tear himself away from his work he enjoyed the society of cultured and intelligent women. He kept up a correspondence with quite a number of such ladies, and had an enviable facility for paying them pretty compliments in several languages. At the same time, he could be critical. In one letter he wrote: 'I personally find the conversation of Parisians the dreariest thing I know, whereas it is delightful to meet cultured and not excessively emancipated Russian ladies. Unfortunately, they have an aversion to soap—but one must not expect too much.'

Apart from his scientific and business activities, much of Nobel's time was taken up by voluminous correspondence and paperwork, every detail of which he coped with entirely alone, from duplicating to the keeping of his private accounts. During his last years he thought at times of engaging a secretary, preferably a woman, but was never able to find one who came up to his expectations.

In the spring of 1876, a certain affection for Vienna and old Austria caused him to advertise as follows: 'A wealthy and highly-educated old gentleman living in Paris seeks to engage a mature lady with language proficiency as secretary and housekeeper.' A reply came from Countess Bertha Kinsky von Chinic und Tettau, who was then thirty-three years old, and a poor relation of the powerful Kinsky family. While employed as a governess to Baron von Suttner's daughters, Bertha had become secretly engaged to the son of the house, Baron Arthur Gundaccar, who was seven years younger than herself. The young man's mother had opposed the romance, partly on account of the age difference, but no doubt owing to Bertha's meagre dowry, and this was why Bertha had applied for a post with Nobel.

In her memoirs, Bertha von Suttner— who later became a champion of peace and a winner of the Nobel prize herself —describes their first meeting, when he received her at the railway station and took her to a hotel where she was to stay until the rooms reserved for her in his elegant mansion in Avenue Malakoff were ready. 'Alfred Nobel made a very good impression on me. He was certainly anything but the "old gentleman" described in the advertisement. He was then 43 years old, of slight, less than medium build, with a dark beard, features that were neither good nor bad, and a melancholy and mild expression in his blue eyes. His voice had a sad but sometimes ironic note; that indeed was his nature. Could it be why Byron was his favourite author?' It is clear from photographs that Bertha herself was then an outstanding beauty. She spoke four languages fluently, was interested in music and literature, and had the qualities of a woman of the world, all of which must have strongly appealed to Nobel. There is no doubt that her personality attracted him, and it is possible that he was even more deeply affected—but his many engagements did not allow them more than two hours a day together, either during lunch in the conservatory at his home, or else driving through the Bois de Boulogne in his elegant carriage.

After only a week's acquaintance Nobel was obliged to go away on business and leave Bertha alone in Paris. Meanwhile, she was deluged by despairing letters from Arthur von Suttner and his younger sisters, her former pupils, who described their brother's misery at her departure. In the end, homesickness got the better of her and, after leaving a note for Nobel explaining matters, she went back to Vienna. Her journey home and its consequences provide a romantic tale; having sold some inherited jewellery to pay for her ticket, she sought out her future husband, who declared that he could not live without her, and they were married secretly in a small suburban church on 12 June 1876, going to the Caucasus for their honeymoon. Not until nine years later did his parents forgive her, and only then was it possible for the couple to return to the family estate.

Countess Bertha's sudden departure must have been a blow to Nobel. He may well have thought of her both as a much-needed secretary and a sympathetic companion who could have given him some feeling of the home life he had missed. They did not meet until eleven years later, when she and her husband came to see him in Paris, but they kept in touch by letter and his high regard for her never changed. During the last ten years of his life he took a keen interest in her projects for peace, and there can be no doubt that she had a strong influence on the final drawing up of his will, and the founding of the peace prize.

Soon after his meeting with Bertha Kinsky, Nobel became friendly with another young Viennese, who was to play an important part in his life during the next eighteen to twenty years, although in an entirely different way. While staying at the small bathing resort of Baden bei Wien, south of the imperial city on the Danube, in the summer or autumn of 1876, he met a handsome twenty-year-old girl with a Jewish background named Sofie Hess, who came from a lower-middle-class family in Vienna. Her father was struggling to support a wife and four daughters by means of a small agency, and in order to help her parents Sofie worked in a florist's shop, where she probably met Nobel. He was struck by her fresh beauty and charm, and they began to talk. He heard her story, became interested and promised to help her to a better position, thus starting a friendship which, for Nobel, soon turned to something warmer. The course of this relationship between two people so basically dissimilar is illustrated in their correspondence. This is preserved in the archives of the Nobel Foundation and comprises two hundred and sixteen letters from Nobel and some forty from Sofie, as well as letters from her father, her sisters and brothers-in-law, lawyers and various other people in Vienna during the 1890s. (How Nobel's letters came to be held by the Foundation is another story, to which I shall return later.)

These documents show Sofie as a pretty, lively and rather irresponsible girl, typical of a certain class in the Vienna of those days, whose aim was to improve its position without too much personal effort, and bask in the reflected glory of the elegant world. Sofie had received a haphazard upbringing, and she seems completely to have neglected the study of French which Nobel arranged for her, and to which he was always encouraging her to apply herself.

It is difficult to understand how a man of such intellect and such high ideals could become infatuated with this untalented and indolent little creature. One reason may be that she at first was so flattered and moved by his attentions that she did everything possible to amuse and entertain him. In his letters he often compared her with a little bird, and her twittering may indeed have helped to counteract his attacks of melancholy—although the incongruity of their relationship was later to become another source of gloom and despondency to him.

In the beginning, Nobel tried to play the part of an elderly friend whose pleasure it was to help a charming young girl with her education and give her a better start in life. He rented a small flat for her in Paris and gave her a generous allowance. But he could not keep up the attitude of benign protector for long. Judging from his letters, Nobel soon fell seriously in love with his young protégée. Yet in this, as in other respects, his behaviour was full of contradictions. In later exchanges with acquaintances, he emphasized that no intimate relationship ever existed between the two of them, though for many years he clearly felt the need of her tenderness and affection. He was often extremely jealous—while at the same time warning Sofie against getting too fond of him, and advising her to think in good time of finding a younger partner.

Apart from their different educational backgrounds and philosophy of life two factors gradually caused serious trouble between them. One was Sofie's insistence on calling herself Madame Nobel, and the other her increasing pretensions and mad extravagance with Nobel's money—though it must be admitted that this was partly his own fault.

The fact that she often posed as Madame Nobel both in Austria and Germany—though not in France—was due at first to their common wish to avoid attention and gossip when they stayed together. During the early half of the 1880s, his letters were in fact often addressed to 'Frau Sofie Nobel'. Not until later, when the relationship had cooled and his business associates and friends—including Baroness von Suttner, who actually wrote to congratulate him on his marriage—noticed Sofie's continued use of his name, did he begin to react against it.

As already mentioned, her extravagance was largely the result of his generosity and infatuation at the beginning of their relationship, when he had spoilt her thoroughly and introduced her to a life of luxury far different from that to which she had been accustomed. Her small but fashionable apartment in the Avenue Victor Hugo was elegantly furnished; she had a personal maid, a cook and, as time went on, even a lady companion who was engaged to teach her French. She seems to have been delicate, and often had to go for cures to spas in Europe. In 1879, Nobel rented a fifteen-room villa for her in Bad Ischl, which he sometimes visited himself. He later bought the house for 22,653 florins and furnished it at great expense, only selling it in 1893.

It is hardly surprising that a woman with no idea either of handling money or keeping accounts now indulged in unlimited spending. In spite of the liberal allowance he provided, Nobel was continually forced to pay her debts and save her from the bailiffs. Yet even though he scolded her severely his patience was truly amazing. The only possible explanation is that he still felt attracted to his *liebes Sofferl*, *liebes süsses Kindchen*—pet names which he employed right up to the final break between them.

During the first few years he tried in various ways to improve her mind and character, hoping perhaps that the two of them would grow closer once the gaps in her education had been filled and certain personal weaknesses eradicated. His perpetual hectoring soon began to bore *das Sofferl*, however, and may account for her calling him *der Brummbär* —Old Growler, a name with which he often signed himself in his letters.

Once Sofie's 'education' had begun to show signs of improving, Nobel apparently decided to take her to Stockholm, to introduce her to his family. On 20 August 1878, he wrote: 'If you are a good girl and get well quickly, I will try to arrange for you to go with me to Stockholm. The main thing is that you are in the best possible health, and I hope for this with all my heart. So, dearest *Sofiecherl*, don't worry your little head about anything but making your Old Growler happy by soon getting well.'

None the less, there was to be no trip to Stockholm for Sofie, either that year or the next, although they discussed it every time Nobel went home for his mother's birthday. Sofie pleaded and raged, even insinuating that he had chosen another woman to take with him, an accusation which infuriated Nobel. His excuses for not allowing her to go were always the same—either her precarious state of health, or his own engagements. It was becoming increasingly difficult to introduce her to his mother, who was so superior to Sofie in every way. He was also beginning to realize how unsuitable she would be as a permanent consort.

One often has the impression that the letters he addressed to Sofie Hess—who, he repeated, was incapable of following his train of thought or understanding his view of the world —were simply a way of salving his conscience, perhaps of easing his mind during certain periods. Yet it is surprising that he often spoke of her to business friends and acquaintances, who had visited him, among them Barbe, Alarik

Liedbeck, Victor Hugo and Walter Runeberg, and had obviously met Sofie. He even introduced her to his brothers, Ludvig and Robert, and was clearly pleased that they were friendly towards her, at least in the beginning. Ludvig, however, who was in some ways more realistic than Alfred, seems to have stressed the drawbacks of their liaison, and advised Alfred to break with her.

Excerpts from Nobel's many letters to Sofie give some idea of their relationship and its progress. It is only after much hesitation that I have decided to publish parts of this extremely intimate correspondence in an attempt to show how strongly the affair with Sofie, added to the events in France and England, affected Nobel's attitude to his closest friends, to his new feelings of warmth for Sweden and, indirectly, to the planning of his will.

The first letter in the collection is undated, but was probably written from Vienna in the early part of 1877. He is describing his negotiations with the directors of his Austrian dynamite company on a subject he had already discussed with Sofie (presumably the manufacture of blasting gelatine in Pressburg):

> Dear, pretty child,
>
> It is past midnight, and we have just finished our third meeting of the day. What keeps us so busy is the old story you know all about. Tomorrow I have to go to Pressburg, where my presence is very necessary. Everything here has been terribly neglected, and if things were not so repellent to me, I would have stayed on. But I am not at ease in the company of the people here, and I long to get away and come back to you. I shall cable or write as soon as possible, telling you where and when we can meet. Until then, a thousand greetings, and every good wish.
>
> Your very affectionate friend, A.N.

Letters two and three, written on 16 and 17 May 1878, and number four from Ardeer in Scotland, are warmer:

Dearest little Sofie,

Here I am, shut up in a wretched cubbyhole of a hotel, with the wind howling from every direction, while my thoughts keep on returning to the delightful time I have just spent in Paris. How are you, dearest child, so far away from your Old Growler? Is your little mind spinning golden dreams of the future—or is your young soul perhaps lingering in the treasure house of memory? Or does a beautiful picture of the present light up your day? Trying vainly to solve this mystery, I whisper from a faraway land my hope that all may be well with you. Good night!

Alfred

In August the same year they had a slight quarrel. Sofie, who was taking the waters at Schwalbach, sent Nobel a cable in French, and in three letters written between 24 and 29 August he wonders jealously who helped her to compose it: 'It doesn't require much wit to guess who put together the telegram you sent me. He doesn't write badly, but even so there are mistakes . . .'

She puts his fears to rest, and in a second letter on 29 August, he writes: 'I realize now that Mademoiselle was at the back of that telegram. Ah, my dear—men are so suspicious!'

Similar strange outbursts recur on both sides during a number of years, though apparently without any basis in fact. Sometimes Nobel would write to her every day, and grow agitated if two days went by without an answer. Yet he kept on advising her not to become too attached to him. A letter written from Stockholm on 27 September 1878 speaks for itself:

My dearest Sofiechen,

No letter from you yesterday. Being so far away this worries me, since we are nearing a difficult time of the year for people of delicate health. I hope you are having good weather; even here in the north, the sun is shining and the weather mild.

Dearest child, you complain that my letters are short and formal—but you refuse to accept the reason for this without forcing me, against my will, to explain it to you. It is because people in general, and especially women, are selfish beings who only think of themselves. I have known from the start and realize more and more clearly as time goes on that your position in life is somehow askew—and I often force myself to appear cold and restrained, so that you will not become too fond of me. You may perhaps believe that you care for me; but this is only gratitude, possibly respect, and such feelings are not enough to satisfy your need for real love. The time may come, and soon, when you will meet someone else—and then you will reproach me if I have bound you irrevocably to myself with ties of deep love. I see this all too clearly . . . though not because of the coldheartedness of which you so often accuse me. Like many people, indeed perhaps more than most, I have to bear the heavy burden of loneliness—and I have long tried to find someone whose heart responded to mine. But this cannot be a twenty-one-year-old heart, with little or nothing in common with my own. Your star is in the ascendant in the firmament of Fate, while mine is declining. In your case, youth glows with all the colours of hope, while with me the few that still blaze are tinged with the rays of sunset. Two such beings can never live together as lovers—but that does not mean that they cannot remain good friends.

What worries me is the thought of your future. If you should fall in love with a young man and he with you, your present false way of life might well be an obstacle to your happiness. I know that you don't in the least care what people think, and this is a blessing which will spare you many a pinprick. But no one is entirely independent of the opinion of others, and self-sufficiency without the respect of one's fellows is like a treasure which cannot bear the full light of day.

Thoughts like these torment me when we are together, and make me irritable when I am with you, sad and pensive when we are apart. I know you are a sweet, warmhearted little girl, and even though you have given—and still give me—a lot of trouble, I am deeply fond of you, and consider your happiness before my own. Did I say my happiness! I must laugh . . . as if this concept were in any way applicable to my nature, which seems to have been created only to suffer. Yet in order to be really happy you still need an education suitable to your station in life, and so you must go on working hard. You are still very much a child, without any thought of the future, and you are lucky to have a kind and considerate 'uncle' to watch over you.

Your letter of the 23rd has just arrived, and I can read between the lines that you are having a good time. It is obvious that your thoughts are far away when you write to me, since there is sometimes as much as half a page without a word of sense! Perhaps this is why you don't get on with Miss B who may not approve of your having too many new men friends. Don't take this as a reproach; I am only trying to find a reasonable explanation for what you tell me.

Enough now of this growling! Only take good care of your health. Don't be tempted by your flirtatiousness to run about too much or to stay up too late at night. And don't force yourself to write long letters to me—it is clear from the last one that you find this tedious.

My sister-in-law is approaching with her children, so I end with a loving kiss from the old philosopher!

September 27

I am staying here until the evening of 1 October. If you cable, the address is Grand Hotel, Stockholm. But write out my Christian name, as my brothers are here too.

How are you off for money?

In the summer of 1880 he was in Hamburg, deeply disturbed by an American court case in which a former German employee named Dittmar was trying to deprive him of his right to the dynamite patent. He wrote to Sofie every day, expressing his longing for her:

My beloved little sweetheart!

Alone here and tormented by unpleasant business matters to such an extent that my nerves are in shreds, I feel more strongly than ever how very dear you are to me. The clamour of the world means less to me than to most people, and I would be happy to retire to some quiet corner where I could live without pretensions, but also free of worry and torment. As soon as the present case has been settled, I am determined to give up business altogether . . . In the meantime, I am hard pressed and have no time to write to you at length, my darling child. I can only say in a few words how I hope with all my heart that you are well, and that the cure will restore you completely to full health.

A thousand warm greetings from your loving

Alfred

A few days later, he urges her not to sacrifice her young life for an impossible liaison with an ageing man like himself. And in December the same year, when obliged to spend a Sunday alone in Glasgow, he sent her one of his interminable 'sermons':

Little Sweetheart!

I should have travelled on to London last night, but the train which brought me here was so late that I missed the connection. And since there are no trains on Sunday in this God-fearing country, I am left high and dry in a hotel as big as a borough.

Alfred Nobel in his thirties.

Alfred Nobel in his fifties.

The blasting of the Corinthian Canal.

Aerial photograph taken with Alfred Nobel's rocket camera.

Left: Alfred Nobel's house in Paris, 59 Avenue Malakoff. Today it is 74 Rue Poincaré.

Below: The conservatory at Avenue Malakoff.

Björkborn, Bofors, The Manor House. Alfred Nobel's last home in Sweden.

Left: The Villa Nobel, San Remo.

Below: Alfred Nobel's laboratory in the grounds of the Villa Nobel.

N:o 73. År 1897 den 5 Februari uppvist vid vittnesförhör inför Stockholms Rådstufvurätts Sjette Afdelning; betygar ex officio.
Jacob Kinders.

Lösen En krona ant. å prot.

Testament

Jag undertecknad Alfred Bernhard Nobel förklarar härmed efter moget betänkande min yttersta vilja i afseende å den egendom jag vid min död kan efterlemna vara följande:

Mina Brorsöner Hjalmar och Ludvig Nobel, söner af min Broder Robert Nobel, erhålla hvardera en summa af Två Hundra Tusen Kronor;

Min Brorson Emmanuel Nobel erhåller Tre Hundra Tusen och min Brorsdotter Mina Nobel Ett Hundra Tusen Kronor;

Min Broder Robert Nobels döttrar Ingeborg och Tyra erhålla hvardera Ett Hundra Tusen Kronor;

Fröken Olga Boettger, för närvarande boende hos Fru Brand, 10 Rue St Florentin i Paris, erhåller Ett Hundra Tusen Francs;

Fru Sofie Kapy von Kapivar, hvars adress är känd af Anglo-Oesterreichische Bank i Wien är berättigad till en lifränta af 6000 Floriner Ö.W. som betalas henne af sagde Bank och hvarföre jag i denna Bank deponerat 150,000 Fl. Ungerska statspapper.

Herr Alarik Liedbeck, boende 26 Sturegatan, Stockholm, erhåller Ett Hundra Tusen Kronor;

Fröken Elise Antun, boende 32 Rue de Lubeck Paris, är berättigad till en lifränta af Två Tusen Fem Hundra Francs. Dessutom innestår hos mig för närvarande Fyratio åtta Tusen Francs henne tillhörigt Kapital som äges att till henne återbetalas;

Herr Alfred Hammond, Waterford, Texas United States, erhåller Tio Tusen Dollars;

Fröknarne Emmy Winkelmann och Marie Win-

Sthlm d. 30 September 1897 den 30 September [illegible]
Abraham Unger.

The first page of Alfred Nobel's Will.

Ragnar Sohlman at the age of twenty-five.

Conscript Sohlman in 1898.

Nowadays, when I am forced to meet others, it comes home to me how badly the lack of social life during the last few years has affected me. I often feel so stupid and ill at ease that I try to avoid the people I meet. This is due to my being such a milksop, indeed I doubt if I will ever recover my spiritual vigour. Don't think I am blaming you, dearest little child—in the long run it is my own fault, not yours. Our attitude to life and its strivings, our need for mental sustenance, and our views of the duties of people of good education and breeding are so utterly different that we can never really begin to understand each other. But I grieve bitterly over the loss of my spiritual nobility, and no longer dare associate with educated people . . .

My reason for making this wretched confession to you is that my heart bleeds with shame at my having become so mentally inferior to other people. Don't be angry with me for such a show of my feelings. You did not realize what you were doing when you took advantage of my compassion and indulgence to undermine my spirit year after year. Alas, such is life; he who pulls back from educated company and gives up exchanging ideas with thinking people, in the long run grows incapable of such exchanges, and loses respect in his own eyes and in those of other people.

Dear, tender little Sofie, I end this letter in the hope that your young life may turn out to be better than mine, and that you will never have to suffer the feelings of degradation which embitter my days. Live happily, and think now and again of your wretched and inconsolable friend

Alfred

In due course, Sofie's prolonged stay in Paris became a trial to both of them. Nobel felt constrained in his work and social life, and Sofie was not happy. She never learned French, and was increasingly homesick for Austria. Having nothing to do made her restless, and this was a further cause for complaint by Nobel. In the next three letters, the first one from 1881, the deterioration in their relationship becomes obvious:

Dear, sweet child,

I have just received your friendly lines of the 28/8. But why do you complain, Sofie dear, that I don't write more warmly? Have I not been telling you for years that feelings cannot be forced! You are a good and lovable girl, but you get on my nerves, and it is part of my freedom-loving nature not to be happy in daily contact with people like you. Especially if they are suspicious, jealous and childish. I admit that for the moment things seem to be better—but even so, I prefer to be alone.

Like you, I take no part in social life, and have scarcely any friends—and like you, I long for the companionship of someone—be it man or woman—who understands me. You do not now, and you never will realize that a free spirit cannot be chained . . .

I used to be very sorry for you, and feel the more so as I get to know you better. If you had made me happy instead of miserable from the beginning, you might have held me . . . Now your young soul is trying to force a love which you rightly find lukewarm. But whose fault is this? I can only repeat: try to win the true love of some good man with whom you can live happily and honestly . . . Your present feelings of emptiness and frustration may well be the cause of your physical weakness. Try to think seriously for once, and keep these well-meaning words of mine in your heart.

Now, no more sermonizing! How long do you think of staying on? The weather is cold, but they say it will improve in a day or two. I sincerely hope so for your sake, as otherwise your present cure will not be of any help.

I spend a lot of time at Sévran and Hamam, and keep to a diet of horseradish and grape juice. Not that it does me much good, but there may be something in it. Anything is better than drifting around spas, with nothing better to do than to kill time. Physical illness is a thousand times preferable to mental stagnation.

Tenderest greetings from your

Alfred

Have you any Tokay left? If not, you can order some bottles from Polngyay.

21.9.83

Dear, sweet little child,

My heart bleeds when I think of you drifting about the world alone—even if it is your own fault. I have been saying for years that you ought to have a lady companion. Your refusal to listen to my suggestion has been the cause of all our suffering. Can't you see what a burden it is for a man as busy as I am to have a woman friend who knows nobody, and who prevents me from moving about freely? This appalling state of affairs has added twenty years to my life. If only you had a companion—obviously a kind and dependable person—I would not be obliged to trot round Europe like a nursemaid, and we could spend more and happier time together. Believe me, our misery is due to your whims and silly childishness. I am not at all difficult to please, and even if I sometimes seem melancholy, I am really not quarrelsome.

Today I sent you a long, calming cable, and I beg you not to let yourself be irritated after your tiring cure.

Whether or not you ought to spend the winter in Montreux is something we can discuss later. It is still warm in Paris, and it will not be cold for some time. Don't you think it rather absurd to have an apartment here? You are away all summer, and now you don't want to come here in the winter either. But you are quite right—you don't suit Paris, and Paris doesn't suit you. Why not choose a place where you would really like to settle down? Montreux, or anywhere else. But to traipse around as you do now, taking up my time until my life turns to gall, is neither wise nor right. Don't you see that I do enough travelling for business reasons, without this everlasting running around with you?

There has been shocking carelessness in Petersburg. Two bills have been protested for Gen. N., not for lack of money but because the funds were sent to the wrong bank. What do you think of that? Don't mention it to anyone, and burn this letter at once.

Ludvig's address is simply Ludvig Nobel, St Petersburg. Mark the envelope 'Personal' otherwise they will open it at the office.

Tenderest greetings, and many kisses from your ever loving

Alfred

11/7/84 Hamburg

My dear little Sofferl,

I see from your cable that the cure at Karlsbad did not do much good—but this is only what I expected. The best cure for you, of course, would be peace and quiet. But you insist on trying to find this in places so far away that I neither can nor want to accompany you there . . . and so it has been for more than seven years, with no advantage to you and constant sacrifices on my part which have embittered and destroyed my life. It is my desire to devote myself entirely to my profession—science—and I regard all women, both young and old, as intruders, who steal my time . . . instead of being able to work in my laboratory, I have been a nursemaid to a grown-up child who thinks she can indulge in any manner of whim. If you were only content to live in some agreeable place in the same country as I do, you could be happy, instead of quarrelling with me in this hopeless fashion, making me the laughing stock of all my friends . . . still, don't let us harp on the past! I only write like this in an attempt to open your eyes. How should we plan the immediate future? You want a villa in Bad Ischl—all right, let us buy one, but what then? I will come there as willingly as to Hell . . . and in due course, Ischl won't suit you any longer, and then there will be complaints, and you will want a new villa in Reichenau or Villach or Görtz or Murzzüschlag—what do I know! . . .

I hope that by the end of the month you will be rested after Karlsbad and that your cure will not be entirely wasted. I myself am so plagued by stomach pains and headaches that you can't expect much of a letter. I am so overworked, both day and night, that I scarcely have time to cable you. It is now two thirty a.m. and the day's conference only just over.

Warmest greetings and embraces from an old friend who is truly sorry for you

Alfred

Hamburg, Thursday night

Sofie moved to Bad Ischl in summer, but could not decide where to stay in winter. She even tried to persuade Nobel to leave Paris and live with her. After her departure from the city in 1884 he was swamped by orders for clothes and trinkets, and for the dispatch of her furniture from the Paris flat. There is something tragi-comic in a long account he sent her in September 1884 of how, in spite of pressing work in his laboratory, important conferences and much correspondence, he had to wander around different fashion shops to carry out her frequently impossible requests:

20/9/1884

My little dove,

I have tried to run all your errands to the best of my ability, though it is almost impossible to decipher your scribbling. 'Pieds', for instance, can only mean 'feet'—and there are no feet for sale at the Louvre! Amputated feet are not articles of fashion in civilized countries. I have ordered a dress at Moret's—light blue rather than navy. But Moret's collection is indifferent, and I would advise you against buying from him. His coats too were much too gaudy. Because of all your moving around I couldn't tell him where to send the dress. At the Louvre I got gloves, laces, veils and scarves in abundance, and addressed them all to Meissl's Hotel in Vienna . . . but I couldn't buy any hats at Reboux, since I don't know where you are . . . The whole thing is a great nuisance, as I am on the point of going away, and can't postpone my journey for the sake of these . . . clothes!

The letter continues with repeated expressions of self-pity and recrimination:

From the beginning I urged you to improve your mind, since it is impossible really to love someone whose ignorance and lack of tact make one feel ashamed all the time. You cannot be aware of your failings in this respect, or you would have done something about them long ago. Even if one were madly in love, the sort of letter you send would act like a cold shower on the heart. It is intolerable that someone who writes as abominably as you do is signing my name in the awful letters she posts round the world! Believe me, my dear, people who have no feeling for culture are only suited to a subordinate place in society, and can only be happy there. You always say that I am incapable of loving; this is untrue, and I could even love you if your lack of education was not a continual torment to me . . . But what is the use of trying to explain; you will never understand that a person can respect such things as dignity and esteem—otherwise we would have agreed on this matter long ago. In any case, I wish you everything good—and hope that you are happier than I am, sitting here sad and lonely. I embrace you tenderly in my thoughts.

Your Alfred

Letters in the same vein—sometimes also accusing her of affairs with other men—continued for the next few years, though often with a lingering note of affection. Writing to her from Paris in March 1886, Nobel apologized at length for having been too busy to remember a cable on her birthday . . . 'but it won't be long until I can come and see you and meanwhile my little girl must be good and show common sense. Did I say common sense, Sofie dear? Now I must really laugh—what is so sweet about you is your complete lack of it! . . .'

Nevertheless, he complained bitterly that it had become impossible for him to go back to Vienna, where he had once been so happy, because of Sofie's use of his name and the gossip this had caused. The last straw, perhaps, was a letter from Bertha von Suttner, congratulating him on his marriage; she had heard in a florist's in the city that a bouquet had been sent to 'Madame Nobel' in Paris, and that a Madame Nobel had stayed in Nice.

In Bertha von Suttner's memoirs, she quotes Nobel's answer:

My dear Baroness and friend!

How ungrateful he is, that fellow Nobel! Though in fact it only appears so, since his feelings for you are as warm as ever, and the closer he comes to the grave, the more he becomes attached to the few people who still show him a little interest.

Do you seriously think I would have got married without telling you? This would indeed have been a double case of lese-majesty, an infringement of Friendship and Courtesy. The old bear has not fallen quite as low as that!

When the florist in Vienna made me out a married

man, she did so with flowers; the Madame Nobel in Nice was, of course, my sister-in-law—hence the explanation of my mysterious and secret marriage. Everything in this mean world can be explained, except possibly the magnetism of hearts, to which that same world owes its continued existence. It would seem that I lack this magnetic quality, since there is no such person as Madame Alfred Nobel—in my case, Cupid's arrows have been inadequately replaced by cannon.

So you see from this that there is no '*jeune femme adorée*'—I quote verbatim—and that in this context I can find no cure for my '*nervosité anormale*' or my melancholy thoughts. A few wonderful days in Harmannsdorf might help me, and if I have not yet replied to your charming invitation, there are a thousand different reasons which I can better explain in person. But no matter what, I must try to visit you soon—if not, who knows whether I will ever have the pleasure and consolation of doing so. Fate, unfortunately, cannot be turned into an insurance company, even if one is prepared to pay the highest premium . . .

My warmest greetings to your husband! I need not repeat that I remain, as ever, your affectionate and fraternal friend,

A. Nobel

Paris, November 6, 1888

It is not known whether any visit to Harmannsdorf actually took place, but he continued writing to Bertha, and met her and her husband in Zurich, where they were his guests for a few days after taking part in a peace congress and an interparliamentary conference in Berne.

It was during these meetings and through the ensuing correspondence between them that Nobel first began to consider the institution of a peace prize. He wrote about this on 7 January 1893 (the letter is copied from his journals, while Bertha von Suttner's memoirs give a slightly different version:

> My dear friend,
>
> A Happy New Year to you and to the courageous campaign against ignorance and stupidity that you are conducting with such energy!
>
> I should like to allot part of my fortune to the formation of a peace fund to be distributed in every period of five years (we may say six times, for if we have failed at the end of thirty years in reforming the present system we shall inevitably revert to barbarism). This prize would be awarded to the man or the woman who had done most to advance the idea of general peace in Europe. I do not refer to disarmament which can be achieved only by very slow degrees. I do not even necessarily refer to compulsory arbitration between the nations, but what I have in view is that we should soon achieve the result—undoubtedly a practical one—that all states should bind themselves absolutely to take action against the first aggressor. Wars will then become impossible, and we should succeed in compelling even the most quarrelsome state either to have recourse to a tribunal, or to remain quiet. If the Triple Alliance instead of comprising three states were to secure the adherence of all, secular peace would be ensured for the world.*
>
> With warmest greetings to yourself and your husband,
>
> Your very affectionate
>
> A. Nobel
>
> *San Remo*, 7/1/92

Alfred Nobel's intention to institute a peace prize was expressed in a modified form in his will of 14 March 1893. According to this document the main residue of his fortune —after deductions for various personal legacies, a gift of

* This paragraph is quoted from Schück's biography.

1 per cent to the Austrian Peace League and 5 per cent to Stockholm's university, Stockholm's hospital and the Caroline Medical Institute respectively—was to be handed over to the Royal Academy of Sciences. This donation was, to quote the will:

> for the purpose of forming a fund, the interest of which shall be distributed by the Academy each year as a reward for the most important and original discoveries or intellectual achievements in the wide field of knowledge and progress, excluding physiology and medicine (where the prizes should be distributed by the Caroline Institute). Although I do not make it an absolute condition it is my wish that such persons should be specially considered who are successful either in word or deed in combating the peculiar prejudices still cherished by peoples and governments against the inauguration of a European peace tribunal. It is my definite wish that all prizes contemplated under my will shall be awarded to the most deserving without any regard to the question whether he be a Swede or foreigner, a man or a woman.*

Returning to Nobel's letters to Bertha von Suttner and Sofie Hess, it must be emphasized that the difference in tone and content is very striking. Those to Bertha are courteous, elegantly-phrased and chiefly concerned with his newly-awakened idealistic interests, while those to Sofie from the middle of the 1880s onwards, when his earlier infatuation had passed, are either a diary of disagreeable events or a dumping ground for moods of depression, containing nothing but reproaches and admonishments. The original tenderness and passion has been replaced by harshness and bitterness.

In a letter of 11 November 1889 he again reproaches her for having lost him all his friends through her behaviour, and mentions that he has destroyed his earlier will:

* Quoted from Schück's biography.

> . . . you say that I could be happy living alone. Alas, this is not true. Illness and fatigue are wearing me out and, before going to sleep at night I often picture a miserable end, probably alone with some old servant who keeps on asking if I have remembered him in my will. The fact is that I have left no will—I have torn up the one I made earlier—and my finances are totally undermined . . . To get rid of my gloomy thoughts, I have begun speculating and this has cost me a fortune. No one who throws money away as I do can have much left at the end of the day . . . but nothing really matters to me any more . . .

The latter part of this outburst is clearly directed at the reader, and it was not until three and a half years later, when he had managed to escape from the circumstances which caused his depression, that he considered making a new will.

As we know, Nobel was often unreasonably jealous, even during his early years with Sofie, but it was probably not until her return to Austria in the late 1880s that she, on being left to herself, began to have affairs with other men. One of them was a Hungarian officer and nobleman named von Kapivar by whom she had a daughter in July 1891. When Nobel first heard of this in the spring of that year, he wrote her a compassionate and generous letter, beginning with the words 'Poor girl':

'What you need now is sympathy rather than recrimination. The blame for what has happened must be attributed to your childhood and early upbringing. Your soul, my dear, is a very small one—but it has never been vicious.'

Nevertheless, he could no longer accept her way of life and increasing extravagance, which was based on the belief that he would always pay her debts. He therefore made it a condition that she should be made a ward of court (declared legally incompetent), promising that if she agreed he would support her with a fixed allowance. After several years of negotiation with various Austrian lawyers it was settled that Nobel would deposit Hungarian government securities to an amount of 300,000 Hungarian crowns (150,000 Austrian florins) in a Viennese bank, thus assuring her an income of 500 florins a month.

In a letter from Aix-les-Bains on 12 September 1894, shortly after his last meeting with Sofie and her child in Vienna, he wrote:

Dear Sofie,

I had to leave suddenly because of a terrible accident in France (an explosion) and there was no time to say goodbye.

You seem in better health than ever before, and I can't see why you complain. Of course things are not perfect, and your present surroundings not as pleasant as you would like. But you are not one of the world's most unhappy people, even if you have done everything imaginable to become so.

It is clear to everyone who knows the circumstances that you have been extremely lucky. Most men in my position would have calmly left you to the misery that you have brought upon yourself . . .

Your small child is charming—see to it that she has the right upbringing. I know nothing of your relationship to her father, and have no means of judging which of you is right or wrong. In any case, it is not my business.

Warm greetings from A. N.

His last letter to her was written on 7 March 1895:

Dear Sofie,

Is it true that your cavalry officer wants to marry you? If so, he is acting correctly, and sensibly as well. You, however, will have to give up much of your conceit, and many stupid ideas. Though when all is said and done, you are a sensitive little creature, and that means a lot. I even believe that you are not wholly without a conscience, as long as Praterestrasse [her family] keeps a hundred miles away.

Warm greetings from A. N.

In 1895 Sofie finally married the father of her child, Captain von K.

Owing to a bizarre coincidence, I later came upon an epilogue to this marriage. While staying in Vienna, I had a car accident, and the Swedish masseur who treated me had spent many years in that city and Budapest. We spoke about Nobel, and it transpired that he knew about the relationship with Sofie Hess, and the break between them. He had even met von K., who told him that after marrying Sofie he had been obliged to leave the army, and had taken a job in a champagne business. He then gave a colourful account of their wedding; having fetched the bride from her home in an elegant carriage, which waited for them outside the church, von K. had led her back to it, kissed her hand, and said good-bye 'for ever'.

The tale was obviously made up. In some begging letters to Nobel, which the latter never answered, von K. asked for an increased allowance for Sofie. But Nobel would have nothing to do with the father of her child, who was later said to have drowned in the Danube.

Despite being a ward of court, Sofie was unable to give up her extravagant habits and continued running up bills, pawning her jewellery and plaguing Nobel with appeals to save her from bankruptcy. Obviously she still counted on being able to arouse his pity, and when he died she was once again heavily in debt. She now applied to the executors of his will for help, through an Austrian lawyer, using both prayers and threats. Her plea was that during their eighteen-year relationship she had been recognized by Nobel as his wife, and that she could prove this in court with a large collection of his letters. If there was no satisfactory legal settlement she would sell these for publication.

Her demand came at a bad moment—1897–98—for Nobel's executors, who were facing problems of many kinds in connection with his will, including the opposition of various relatives, lack of co-operation from the prize-giving academies, and a succession of lawsuits both in Sweden and abroad. They were therefore anxious to avoid further trouble, especially anything which might reflect badly on Nobel himself—and there was no knowing what scandal might be brought to light by the publication of the letters. After lengthy discussions with the German lawyers involved in the work on the estate, who advised a friendly settlement, and following a considerable reduction of Sofie's first claims, an agreement was reached whereby she handed over 216 original letters and one telegram from Nobel, a portrait of him, and an affidavit that she had no further claims apart from the annual income he had designated for her. Should it later transpire that she had kept back letters, or if she tried in any way to injure Nobel's reputation, the executors would be entitled to stop the payment of interest on her capital. In return for this settlement, they agreed to pay her accumulated debts of 12,000 florins. For safety's sake, an Austrian lawyer was appointed trustee for the duration of the negotiations. After the constitution of the Board of the Nobel Foundation, this trusteeship was rendered null and void.

The above account attempts to explain the different circumstances which combined to produce Nobel's strongly depressive state during the early 1890s. The outward signs of this are obvious if one compares photographs taken when he was a young man with those taken at forty-five and fifty-eight, and finally those from his last year, when he had succeeded in almost completely overcoming his depression, and was again full of energy and future plans, despite his increasing heart trouble.

The Legacy

Immediate measures in San Remo

With regard to the situation in San Remo in December 1896 I must stress that none of the relatives and friends who had gathered at Nobel's deathbed knew anything about his will. Personally, I was in a state of shock, being deeply saddened by the unexpected death of a close friend and benefactor and at the same time tormented by the thought that his last hours had been spent alone among strangers who were unable to understand what he was trying to say.

Uncertainty about the future of the staff in the laboratory and the experiments with which I had been entrusted added further to my distress.

On 12 December my wife wrote from Bofors: 'Sederholm and Feilitzen [two staff engineers] came to see me this evening. They clearly feel very anxious about the immediate future, but all we can do is to wait until you come home and tell us about Nobel's plans and what is to be done with the laboratory . . . how sad for you to have lost such a good friend so early . . . but you will always have the memories of your time together.'

After I had gone to bed at my hotel on 15 December, Emanuel and Hjalmar Nobel arrived to inform me that a cable had been received from Stockholm with the news that Nobel's will—which was deposited in a Swedish bank—had been duly opened and that Rudolf Lilljequist and I had been appointed the executors. No further details were available, bar the directive that the veins of the deceased were to be opened and competent doctors called in to sign the death certificate.* The body was then to be cremated.

The announcement of the unexpected responsibility conferred on me gave a sleepless night. I was now faced with a task for which I felt totally unqualified, and would have to work together with Lilljequist, who was a complete stranger to me. Since Nobel must have realized how little I knew of such matters the only explanation for his choice of me as an executor was that he trusted me to carry out his wishes. I finally decided that I must do this, to the best of my ability.

The first step was to prepare an inventory of his property and personal belongings. With the exception of the servants' quarters all the rooms in the villa had been sealed off by the Swedish Consul in San Remo, and I was therefore now obliged to break the seals and take over the keys to the house and the laboratory.

I cabled Lilljequist asking for authority to act on his behalf and on the following day received the answer: 'Do not understand your telegram. Am I mentioned in the will?' Obviously he, too, had been taken by surprise and since he did not know me he cabled Consul Marsoglia to represent him.

A day or two after these events a copy of the will finally arrived in San Remo. It had a depressing effect on all of us, particularly Emanuel Nobel. On going through Nobel's papers we had found an earlier will on which the deceased had written: 'Rescinded and replaced by a later will, drawn up on 27 November 1895.' In the new document sweeping reductions were made in the family legacies, including that of Emanuel, and he was no longer named as an executor.

* This directive was based on Nobel's fear of being buried alive, a terror he had inherited from his father. The Italian doctor who looked after Nobel during his last illness, and who signed the death certificate, was not a little shocked by this request and pointed out that the opening of the veins had already been performed during the embalming process. At that time a large number of poisonous substances had in fact been injected, among them ethyl nitrate (which in certain circumstances becomes explosive).

Compared with the former will, the new one contained many formal discrepancies; the chief beneficiary, i.e. the future fund, was non-existent and would now have to be constituted. It was doubtful how and when this could be done and what would happen during the interim, especially as regards the Nobel Brothers Naphtha Company, which Emanuel Nobel represented and in which Alfred Nobel's holdings had a decisive influence.

From the very beginning Emanuel was averse to criticizing his uncle's will or acting against the latter's express wishes.

He often spoke to me about his worries in this connection, and I am always touched when I recall his friendliness and generosity towards me, who had been appointed to represent interests seemingly opposed to those of the Nobel family. During the difficult months ahead he was to give me the strongest support, and I particularly remember his remark: 'Always keep in mind the meaning of the Russian word for an executor: "*Dusje Prikaztjik*—the vicar of the soul"—and try to act accordingly.'

I received an admonition of a different kind when told that a representative of Nobel's interests, particularly the Naphtha Company, could not afford to live as unpretentiously as I had been doing so far. I would now be obliged to put up at the best hotels in San Remo, Paris and elsewhere, and instal a special reception room—in other words, I was expected to adopt much the same standard as a director in the Naphtha Company.

My main worry, however, was connected with the future of Bofors without the impact of Nobel's personality, although I had understood him to say that the work there would carry on in some form or other after his death. The thought of having to disrupt or sell his far-flung interests in the various enterprises—Bofors, the Nobel Dynamite Trust Company and the Naphtha Company—was truly daunting. In my first letter to Lilljequist on 23 December 1896 I proposed that these investments should be set aside, at least for the time being, as assets for the fund:

I have racked my brains to interpret the clause in the will which says that 'the executors shall convert the residue of my property into cash, which they shall then invest in safe securities'. If this means that all industrial shares must be converted into government bonds or other safe securities it would be a disaster for every single one of the Nobel enterprises. In view of our knowledge of Nobel's own aims and activities, this simply cannot have been his intention. Considering his unique financial genius his own decisions regarding the investment of capital should surely be upheld as far as possible—it would certainly be difficult for anyone else to make a better selection.

Another important consideration is what to do with the various inventions and patents which are either completed or in preparation. And what is to happen about the engineers in Nobel's laboratory, who cannot simply be dismissed? I still hope that some sort of directive will turn up among his papers, since he often told me that he wished that work on his ideas should be continued after his death. In the meantime I have taken upon myself to promise them that all work will continue as usual for the time being. I sincerely hope you will agree to this, at least for a year, during the winding up of the estate.

I have to admit that at this stage my own view of my work as an executor was blurred by my emotions. Furthermore, the immediate administrative problems blinded me to the magnanimity and foresight expressed in Nobel's will and it was not until later that I better understood his thoughts and could devote myself more wholeheartedly to carrying out his instructions. Fortunately Lilljequist had a more realistic and less personal approach, and thus came to exercise a healthy and restraining influence on our future decisions and actions.

After making arrangements regarding the villa in San Remo, Emanuel and I left for Sweden. I myself spent

Christmas in Bofors, where I was able to consult the new manager, Commander Gerhard Dyrssen, and shortly after I met Rudolf Lilljequist in Stockholm.

He was about fourteen years older than me, and considerably more experienced in financial matters, as well as being of a more critical turn of mind. From the very beginning we got on extremely well.

Since neither of us knew much about the formalities we were faced with as executors it was of great importance to engage a Swedish adviser to wind up the estate. Our choice fell on Carl Lindhagen, a justice in the Court of Appeal. Lindhagen was a skilful and broadminded lawyer who showed great interest in implementing the concepts of Alfred Nobel, and was able to establish good relations with the prize-giving academies as well as with the government authorities concerned.

He worked with us for nearly four years and later helped to draw up the main statutes of the Nobel Foundation.

The attitude of the Swedish press to the will

On 2 January 1897 a Swedish newspaper published some of the most important provisions of Nobel's will, chief among them the establishment of a fund for the annual awarding of prizes within five areas of human progress. Both at home and abroad this news was greeted as sensational, and the benefaction described by the paper as 'a gift intended to further the progress of humanity and serve lofty purposes, indeed perhaps the greatest donation that any man hitherto has possessed the will or the means to provide'.

Another paper wrote: 'In Swedish history only one other action can be compared with this one, namely the donation of his private inheritance made by King Gustavus Adolphus, who thereby ensured for all time the existence and development of our chief university, and thus the future of culture in our country.'

The publication came earlier than the executors could have wished. We had hoped to find further documentation either in Paris or San Remo which would help to clarify Nobel's intentions. Apart from its lack of precision, the will also exhibited formal flaws, among them the fact that the chief beneficiary, namely the Nobel Fund, did not yet exist. A very responsible task had been assigned to the institutions appointed to award the prizes; this would entail much work, without any prescribed remuneration, or indeed instruction as to procedure in cases where no suitable candidate could be found.

Finally, there was the legal difficulty of establishing Nobel's true domicile at the time of his death. This was of vital importance, since it would decide which court of law should be recognized as competent to settle the winding up of his estate.

Critics had by now begun to question the very basis of Nobel's will and to emphasize and exaggerate its formal flaws. There were several attacks in the press, clearly supported by Nobel's relations in Sweden who had been advised to contest the will and to aim at a compromise through which the estate would be divided between the nearest heirs and the Swedish institutions mentioned by Nobel.

Among the grievances aired was Nobel's supposed lack of patriotism in bypassing Swedish interests in favour of international activities, and the difficulties which might face the prize-giving bodies in accomplishing their task—it was thus suggested that it would interfere with their main functions and expose their members to bribery and corruption. The awarding of the peace prize through a committee appointed by the Norwegian Parliament was also criticized, since this provision was seen as affecting Sweden's political interests and her relationship to Norway. These attacks clearly influenced members of the prize-giving bodies, and they now began to doubt the wisdom of having anything to do with the matter until the validity of the will was established.

Meanwhile, both Conservative and Left-wing papers continued to question it, stating that it was neither suitable nor practical to accept the provisions. A long letter signed by the leader of the Social Democratic party, Hjalmar Branting, and headed 'Alfred Nobel's will—Lofty Intentions, Lofty Blunders', sharply attacked Nobel for his lack of feeling for the social issues involved—'the products of nature and hard work should be available to all'—and chastised him for placing 'these great prizes at a point so far ahead on the human race track that those who manage to reach it already enjoy most of what our society can offer in material and spiritual terms'.

Referring to the awarding of the literary prize to 'the person who shall have produced the most distinguished work of an idealistic tendency' Branting continued: 'It must be openly said that the entire donation is botched by the unfortunate choice of the Swedish Academy as prize distributor —a mistake accentuated by the interpretation this body can be expected to give to the expression "idealistic"'. The peace prize too was criticized, Branting, one of the future laureates, declaring that 'the only way goes through an international union of the working classes in every country . . . Effective attempts to bring about world peace are never the work of a single individual, and it is obvious that the masses should share in the proceeds of the Nobel fund in order to be able to continue and intensify the work for peace.'

Indicating Marxist tenets, Branting finally challenged not only the donation but the capitalist system which made it possible. 'A millionaire who makes a gift of this kind may personally be worthy of respect—but we are better off without either the millions or the donations,' he concluded.

The question of Nobel's domicile at the time of his death

One clause in the will which could not be contested was the appointment of the executors; even so we had to work hard to maintain our position, one difficulty being that the powers of an executor at that time were not recognized in Swedish law, but simply based on legal usage.

The main problems facing us came under three main headings:

1. Legal formalities and controversial issues.

2. The conversion of shares and their re-investment in safe securities.

3. The establishment of the necessary institutions to administer property and rules for the distribution of prizes.

The legal questions comprised on the one hand the formal procedure regarding property abroad and on the other a number of complex legal issues. The former included probate, the rights of the executors, the preparation of inventories, the recovery of accrued interest and measures for the prescribed sale of shares.

One of the most controversial issues was the choice of court to deal with probate.

At his death Nobel owned magnificent houses in Paris and San Remo, as well as a mansion at Björkborn at Bofors in Sweden. But since leaving Sweden for Russia at the age of nine he had not been registered as a resident anywhere and had been jokingly called 'the richest vagabond in Europe'. Formally speaking, the court competent to handle the estate would be in his last place of residence, namely Stockholm. Acting for the executors, Lindhagen therefore now applied for probate in the Stockholm City Court.

Owing to his work at the new electrochemical factory at Bengtsfors in northern Sweden and the difficulty of cross-country travel in those days, Lilljequist could not get away to Stockholm for more than a few days at a time, and it thus fell on me to undertake the journeys abroad necessitated by Nobel's interests in Europe.

My first trip to Oslo in early 1897 was intended to discover the attitude of the Norwegian Parliament (Storting) to the will, and I now met the President, Sievert Nielsen, and other leading members, who all seemed greatly interested in the task assigned to them.

In the middle of January the same year I, my wife and our accountant Seligman set out for Paris, where we stayed at the

Hotel d'Albe on the Champs Elysées, an old-fashioned building close to Nobel's house in Avenue Malakoff. I had been advised to contact the Swedish Consul General, Gustaf Nordling, for instructions how to proceed with the inventory of Nobel's property in France. He was most friendly and helpful and put me in touch with a lawyer, Paul Coulet, who was attached to the Appeals Tribunal in Paris and did work for both the Swedish Consulate and the Legation.

From the very beginning I realized that Nobel's domicile would in French law probably be established as being in Paris. The validity of the will could in that case be called in question in France, where its formal deficiencies would be highlighted by French lawyers applying the strict and detailed rules of the Code Napoleon. If Paris were seen as his *domicile de fait* the French courts would claim a say with regard to property remaining in France and possibly also in other countries, particularly in Germany. In addition, his possessions in France, including foreign securities, would be liable to French inheritance tax. If, on the contrary, Sweden was proved to be his 'true' domicile, only French valuables would be subject to French taxation. The French authorities were therefore likely to do everything possible to have Paris declared Nobel's *domicile de fait*. From every viewpoint it was therefore of the utmost importance to have his last permanent residence accepted as being in Sweden and the competency of the Stockholm court to handle the estate thus established. It was clearly going to be extremely difficult to convince the French authorities on this point; the only chance of doing so was to maintain that Bofors rather than Stockholm represented such a domicile and that the Karlskoga court was therefore the proper tribunal for all matters concerning the will.

I thus wrote to Lindhagen urging that the competency be transferred from Stockholm to Karlskoga, and in due course the Karlskoga court did in fact handle probate.

An official document was required to allow the executors to dispose of Nobel's property in France. This was to confirm their authority according to Swedish law and thus posed quite a problem of wording for Lindhagen. Finally, Nordling came up with a so-called *Certificat de Coutume* in the correct French legal terminology, describing the rights of executors in Sweden.

One paragraph in this document, which admittedly worried Nordling and me, stated that the executors could make an inventory in Paris without actually inviting the heirs to be present. I had written to Lindhagen pointing out the delays and inconvenience such an assembly would involve and he had replied advising against it unless specifically demanded in French law. In Sweden it was thus only necessary to summon relations during the final inventory proceedings.

On arriving in Paris Lindhagen declared Nordling's certificate to be 'excellently formulated and correct in content'. Nordling was, however, severely criticized by the heirs, who also attacked the executors for the use of 'false documentation'.

After a notary had carried out an inventory of the furniture and personal effects in the Avenue Malakoff house in the presence of Nordling and myself and a list had been drawn up of the securities deposited in different banks in Paris, I left to attend to similar business in San Remo, where I was assisted by our accountant and Nobel's own aides.

Three weeks later I was called back to Paris by a letter from Emanuel Nobel and a telegram from Nordling, urging a meeting with them both as soon as possible.

It appeared that the publication of Nobel's will had placed the Naphtha Company in serious difficulties and that Emanuel himself was now in an awkward position. Rumours of an immediate sale of securities, spread by competitors, had caused the company's shares to drop. In addition certain business transactions he had planned with Nobel had fallen through and Emanuel was now being strongly urged from several quarters to contest the will. He himself was not sure how to proceed. I advised him to wait and see, promising that I would try to prevent any forced sale of the Baku shares and seek a compromise in the matter.

Nordling now arranged a meeting in Paris with a well-known Swedish resident, Per Lamm, and a prominent French journalist, Adolphe Brisson. Soon afterwards Brisson published several articles in *Le Temps* repudiating some unfavourable comments on Alfred Nobel which had appeared in the French press. This pleased Emanuel, who himself put forward some good points about his uncle and commented in a positive manner on the intentions expressed in the will. He implied, however, that we could expect opposition from several of Nobel's relations in Sweden.

My own position was decidedly unpleasant. The relatives likely to oppose the will were firstly Hjalmar and Ludvig, sons of Alfred's elder brother Robert, then Count von Frischen Ridderstolpe, representing his wife, their sister Ingeborg, also Professor Hjalmar Sjögren who was married to Emanuel's sister, and finally Mr Åke Sjögren, stepfather and guardian of the three daughters of Emanuel's deceased brother Carl.

Ludvig Nobel and I had been childhood friends and fellow students at the university. I had also come to know Hjalmar well when he stayed at Björkborn, but had only met the others briefly.

It was clearly going to be extremely problematical to uphold the terms of the will and yet avoid a clash with the family. But I was determined to try.

After exchanging some letters with Ludvig and Hjalmar I wrote to Ludvig on 20 February 1897 in connection with the press debate on the validity of the will:

> . . . You are aware, of course, that several Swedish papers have openly advocated a complete dismissal of the will and, in this context, have made perfidious personal attacks on your uncle.
>
> May I say that if we were only dealing with Emanuel,

Hjalmar and yourself the whole situation would be very different. Personally I wish for nothing better than for the executors to continue friendly contacts with the family and I will do everything possible to achieve this. At the same time I have to keep in mind the importance of not compromising the trust placed in me . . .

Please believe, Ludvig, that if I could get out of this business in an honourable way I would do so. It is seriously interfering with my life and work. As things stand I now have to act as a kind of buffer between many different interests and people who are at loggerheads with each other.

Still, this cannot be helped, and I would be a traitor if I did not give priority to your uncle's wishes. Emanuel has promised to remain my friend even if we may differ on official matters. Can I hope that you will do the same?

Your affectionate friend
Ragnar Sohlman

I quote this letter partly to justify measures which I was obliged to take later. His reply was amicable but guarded, and a certain coolness developed between us.

In view of the way matters were threatening to turn out, Nordling and Coulet decided to speed up the liquidation of the estate in France and thus avoid the risk of a lawsuit in a French court on the subject of competence and the legality of the will. Being unable to tackle this problem alone I asked Lilljequist to give Nordling power of attorney if he himself was prevented from coming to Paris. Meanwhile I wrote to Lindhagen suggesting that he should make his way there as soon as possible.

Nordling's power of attorney caused a slight difference between us—the only one, as it happened, in our long association. After consulting with me Nordling had sent Lilljequist an extremely complex formula in French for signature. This document covered four closely-written pages containing every imaginable contingency with which Nordling might have to deal. This annoyed Lilljequist, who took it as an attempt to undermine his authority as executor, and he returned the paper with a stiff note to the effect that he had no intention of issuing such a power of attorney except possibly to me. Nordling took great offence at this, and I myself was reprimanded by the Swedish–Norwegian Minister, Due, for lack of respect towards His Majesty's Consul General in Paris.

Lilljequist finally arrived in Paris, and after informing himself of the situation and getting a taste of French legal procedure, signed the necessary power of attorney for Nordling.

We now decided to withdraw the securities deposited by Nobel in various French banks, in case of seizure by the relations. On the strength of Nordling's *certificat* we thus successfully removed all documents deposited in Nobel's name, placing them in three large strongboxes in the *Comptoir d'Escompte*, pending their removal from France to a safer place, an operation which lasted several weeks.

In the beginning of March I left for London, where I contacted a Scottish lawyer, Mr Timothy Warren, who was to act as our adviser and solicitor in regard to Nobel's estate in England and Scotland. Mr Warren, who had been recommended by the head of the Nobel Dynamite Trust Company in Glasgow, was a typical mixture of Scottish dourness and dry humour. A distinguished man of business and senior partner in a well-known legal firm, Messrs Moncrieff, Barr, Peterson and Company, he was often consulted by the Dynamite Trust. He had known Nobel personally for some years in connection with the cordite case, and his assistance in the winding up of the estate in Britain was to be invaluable.

I then called on the head of Nobel's London bank, the Union Bank of Scotland, to arrange for a later transfer of securities from Paris to London for safe custody and future sale.

Finally, I contacted the directors of an English shipyard,

Palmer & Co. in Newcastle and Jarrow-on-Tyne, who had made inquiries through an Anglo–Swedish businessman as to a possible future merger between the shipyard, Bofors, and an English manufacturer of armour plating, Brawa & Co. More about this later.

A comparison between the two wills

Both wills had been written out by Nobel himself, without any legal help, and witnessed in the Swedish Club in Paris, where two of the four witnesses had signed both documents. The first one, with the words 'rescinded and replaced by my will of 27 November 1895' was found among Nobel's papers in San Remo, while the second had been deposited in the Trust Department of Stockholm's Private Bank (Enskilda Banken), probably in June 1896.

According to the earlier will the legacies, in the form of a certain percentage of the residue of the estate, were to be distributed among twenty-three private individuals, six of them nephews and nieces. This part totalled 20 per cent, while Stockholm University, the Stockholm Hospital and the Caroline Institute were to receive 5 per cent each and the Austrian Peace League 1 per cent. The remainder was to go to the Academy of Sciences in Stockholm to form a fund, the interest on which should be distributed annually, in accordance with directives stated earlier.

The will also included instructions as to the disposal of Nobel's homes—the house in Paris, the country houses in San Remo and Bad Ischl—as well as royalties from patents. The latter were to be used to build crematoria in the larger Swedish cities, the Caroline Institute being directed to take charge of this task.

There were significant differences between the two wills. Assuming the net value of the estate to be thirty million crowns in liquid assets, Nobel's relatives were, according to the first will, to have received some 2.7 million, plus houses valued at half a million. In the second they were left only one million, in carefully specified legacies. Other legacies were reduced in the same manner, some abandoned and some transformed into annuities.

Among institutions favoured in the first will, Stockholm University, the Stockholm Hospital and the Austrian society were left out. The withdrawal of the legacy to the university may have been due to the internal bickering which Nobel knew was going on at the time, or simply to the fact that he considered the university solely as an educational centre and preferred to support research. The Stockholm Hospital he probably regarded as the responsibility of the local municipality; the peace movement, on the contrary, was very generously treated in the second will.

Nobel's reasons for reducing the legacies are partly explained by remarks he made to the witnesses who signed his final will, and are reported in the minutes of the probate proceedings.

From a purely formal viewpoint the earlier will had the advantage that the chief beneficiaries were already existing scientific and humanitarian bodies and that there could therefore be no doubt as to the validity of the will on this point. In addition, the very nature of those beneficiaries was a sufficient guarantee that the future income from the donation would be free of tax.

Apart from any consideration of the personal legacies, on which it is clearly not my business to comment, I regard it as fortunate that the second will, despite its formal flaws, became the main instrument for the implementation of Nobel's intentions. It should certainly be stressed that the origin of the Nobel Foundation, which has come to mean so much for Swedish culture, is to be found in the provisions of this will.

Another advantage of the latter will was the distinction made between scientific and literary activity on the one hand and the peace movement on the other. The task imposed on the Academy of Sciences by the earlier will—the choice between all the various discoveries and achievements in several fields—would assuredly have been less acceptable to world opinion than the selection procedure outlined in the second will.

In the first document the winding up of the estate was entrusted to three executors, Nobel's nephew Emanuel, Consul General Nordling and the director of the German Dynamite Company, Max Philipp. The reason for Philipp's exclusion in the second will was possibly Nobel's belief that he did not know enough about conditions in Sweden. As regards the exclusion of his nephew, Nobel may have realized that the work of an executor might put him in a false position with his family, and that it would be wiser to give the task to two independent Swedish citizens such as Lilljequist and myself.

Opposition among the relatives

After returning to Paris to arrange for the transfer of the main part of Nobel's assets from the safe deposit in the *Comptoir Nationale d'Escompte* to a strongroom in the same bank (kept in the name of Nordling and the executors) I went back to Sweden.

Probate had been issued at Stockholm City Court on 5 February and at Karlskoga District Court on 9 February. At the City Court two of the witnesses of the final will reported on oath the remarks that Nobel had made in their presence about his reasons for changing his will.

The publication of these depositions and a statement by one of Nobel's closest friends, Charles Waern, on the same topic, caused a sensation and started a fresh press debate about the implementation of the will. Parts of this material also irritated Nobel's relations in Sweden, particularly Robert Nobel's family who, not without reason, felt themselves pilloried by Nobel's criticism of his deceased brother's will.

Meanwhile, discussions in Stockholm between Lindhagen and the executors resulted in the dispatch of the following letter to the prize-giving bodies:

We, the undersigned, who have been appointed executors of Alfred Nobel's estate, herewith [beg to] present a duly certified copy of the said will, with a humble request that . . . the institutions in question will accept the task of awarding the prizes . . .

It is, however, clear that the provisions concerning the method of awarding the prizes require further elucidation before they can be applied as prescribed by Dr Nobel. It is our belief that in the case of prizes to be distributed in Sweden, this matter could be arranged by appointing delegates from the various bodies—two from the Academy of Sciences and one each from the Caroline Institute and the Swedish Academy—to negotiate with the executors on the subject. A definite proposal can then be adopted, signed by the bodies concerned and submitted to the Minister of Education with a humble request for the approval of the King in Council; in this connection it is our intention to present to the King in Council the necessary directives for the administration of the fund.

Stockholm, 24 March 1897
Ragnar Sohlman Rudolf Lilljequist

A letter was also dispatched on the same day to the Norwegian Parliament, and on 26 April this body agreed to accept the assignment Nobel had given them.

Several people now advised us to present the king with a copy of the will. Lilljequist and I therefore attended a public audience and were cordially received by His Majesty, who, however, made no comment at the time either about the will or its implementation.

'The Coup'

Telegrams and letters from Nordling—informing me that Hjalmar, Ludvig and Ridderstolpe had arrived in Paris to study the possibility of bringing an action—made me hurry back to the French capital. There we at once set in motion the transfer of valuable documents to London and Stockholm. All shares and bonds considered convertible were thus dispatched to the London office of the Union Bank of Scotland, while government and other securities to be retained in the portfolio of the estate were transferred to the Private Bank (Enskilda Banken) in Stockholm. The papers were dispatched in parcels by registered post, and handed in at a special *Expedition de Finance* at the Gare du Nord. A suggestion that I myself should make several journeys to England and Sweden with them was dropped as too risky and time-consuming. Since the French post office was not able to insure for more than 20,000 francs, we now asked the famous banking house of Rothschild to insure the parcels. The bank stipulated that the highest value of the daily transfers should not exceed two and a half million francs, and that each must be backed by a French insurance policy of 20,000 francs.

The documents were fetched from the strongroom in the *Comptoir Nationale d'Escompte* and taken to the Swedish Consulate to be listed, packed in bundles and sealed before being taken to the station. The actual transport from the consulate was naturally risky, and strict precautions were necessary to avoid suspicion.

After packing the papers in a suitcase at the bank, Nordling and I took a horse-drawn cab to the consulate. I sat with a revolver at the ready in case of a direct attack or a prearranged collision with another vehicle, a trick not unusual among thieves in Paris at the time. The same method of transport was employed for our visits to the Gare du Nord.

Today it may seem odd that we made these convoluted arrangements instead of simply instructing the bank to organize the transfers. The reason, of course, was that we were afraid of calling attention to the move and being stopped by the local authorities, since the question of French tax on the valuables was still unsettled.

One morning, Hjalmar, Ludvig and Ridderstolpe paid a

visit to Nordling's office to make inquiries about the validity of the will. They were not told that I and my accountant were busy in the next room arranging papers, and Nordling now began to feel distinctly uneasy about the situation. After the last batch had been dispatched he therefore urged that Hjalmar and Ludvig must be acquainted with what had happened, and suggested that we should all smoke a pipe of peace over an informal dinner, during which I would break the news about the French securities. I could not, of course, object and a splendid supper was arranged at the well-known Noel Peter restaurant in La Passage des Princes.

The atmosphere, which was strained at first, improved noticeably as the meal progressed, and it was not until coffee was served that the well-worn question of Nobel's *domicile de fait* came up. Hjalmar tried to corner me by stating that the claim about Bofors being his true domicile was entirely fictitious and that we all knew it to have been Paris or possibly San Remo— though preferably Paris, where he had lived for seventeen years and still had a house and a staff. The competent court was thus obviously a French one and such a court, he contended, should also determine the validity of the will.

Replying that the matter was certainly open to discussion I pointed out, however, that it was now of purely academic interest, since all important papers had already been removed from France and were therefore no longer subject to French jurisdiction. I advised him to give up any thought of a lawsuit in France and, if he still wished to proceed, to take up the matter in Sweden.

This announcement acted like a bombshell and Hjalmar would not believe me at first, although my statement was confirmed by Nordling. He then suggested rather vaguely that we should submit our differences to what he called *arbitrage*—a conciliation board.

Later the same evening I left Paris for San Remo with two large suitcases full of letters and papers from the Malakoff Avenue house, which was now up for sale. By this time I was really feeling exhausted, and let Nobel's butler look after the registration of the baggage.

Arriving in Nice the following day I was surprised to be met by the manager of the Nobel Dynamite Trust Company, Max Philipp, who joined me on the onward journey to Monte Carlo. Hjalmar had apparently cabled asking him to catch me up, point out how unsuitably I was behaving, and arrange for a formal agreement through *arbitrage*. Philipp used every conceivable argument to convince me that I was risking my entire future by my ill-advised actions, instead of leaving the whole business to lawyers and judges in the different countries concerned. We ought to take our time in the matter, he declared, and thus be better able to transform Nobel's Utopian dream into something more sensible.

I told him at once that I was acting on my own responsibility, to ensure that Nobel's last will was respected, but he shrugged this off, insisting on a solution of the problem through *arbitrage*—a proposal which I naturally turned down.

My irritation over all this interference had hardly subsided when I arrived at the Customs in Ventimiglia only to find that the two suitcases registered in Paris were in fact not on the train. I waited impatiently but in vain for the next train, and finally in desperation went on to San Remo feeling that I had really put my foot in it if all Nobel's private papers and accounts as well as his personal belongings had been lost through what might well be conceived as deliberate carelessness. The next twenty-four hours were nerve-racking and the tension only eased when a telegram from Paris announced that the luggage had been sent via Modena and would arrive late.

Looking back on the transfer of Nobel's French bank deposits to Sweden and England—even though it was backed by the Consul General, Lindhagen and our French lawyer—I must admit that it may well appear as a dubious measure. I would therefore like to explain the reasons behind the operation in more detail.

The question of competence was still, as we knew, unsettled in France and in the French legal view there were clearly powerful arguments for declaring Paris as Nobel's *domicile de fait*. This would mean, however, that additionally a French court might query the validity of the will. In view of the rigidity of the Code Napoleon the implementation of Nobel's ideas might thus be seriously jeopardized.

It was obvious to me that Nobel himself had never envisaged this possibility. His interests might have been international, but when making his will and writing it out in what he evidently believed to be Swedish judicial language he had clearly taken it for granted that probate would be granted by a Swedish court.

We, the executors, believed that this could be achieved, and were determined to do everything possible to bring it about.

It should be stressed at this point that Nobel frequently switched his securities from one bank to another, sometimes between different countries, presumably in order to keep a check on them. Had it ever occurred to him that Paris might be established as his permanent residence, he would undoubtedly have deposited them in Sweden, as we now had done.

In the hope of softening the hostile attitude of some of the Nobel relatives and of getting a better idea of what they were driving at, I wrote to Hjalmar from San Remo:

> Dear Hjalmar—when we last parted you suggested that 'this will business be submitted to *arbitrage*'. On my way here I ran into Philipp, whom you had asked to approach me on the subject . . .
>
> May I now give you my own view of the situation, which is also that taken officially by the executors.
>
> To begin with, neither Lilljequist nor I have any axe to grind in the matter—we have simply been appointed to look after the interests of others under the terms of the will and to interpret the will as Nobel intended. All

our actions so far have been taken in accordance with these principles.

To me arbitration would only seem possible on points in which the will is either ambiguous or else gives no directives at all. We consider ourselves unable in a judicial sense to enter into any sort of negotiation about the will itself, and maintain that it is our moral duty to defend it as best we can.

As you probably know, I have had long talks with Emanuel about the will and his own interest in it, and he seems satisfied with the results. I only wish I could have started off in the same friendly way with you all—I believe that you and I and Emanuel together might have worked things out. But I soon realized that this was not to be . . .

Any hope of a settlement now lies in your deciding jointly on what you wish to submit to *arbitrage* and instructing your lawyers to start preliminary negotiations with the executors . . .

Sincerely, Ragnar Sohlman

Though the letter gave no immediate result, it may be seen as a first step towards the negotiations which a year and a half later resulted in a friendly settlement with the heirs.

Our dealings in Paris led to an immediate increase in the activity of Nobel's relations there. It was rumoured that the famous French lawyers they had consulted were highly embarrassed at not having realized what was happening and taken steps to safeguard the securities, and on their advice the family, headed by Hjalmar, now demanded the sequestration of all property in Paris belonging to the estate. This chiefly affected the house in Avenue Malakoff which hence could not be sold and thus caused a temporary loss. Hjalmar also travelled to Germany to ensure the sequestration of Nobel's property there. I had already employed a well-known lawyer recommended by Dynamite A.G. to deal with these assets, totalling some six million marks which were deposited in German banks and companies. The appointment of executors was not, however, immediately accepted under German law, and Lindhagen was therefore obliged to make out a certificate similar to that which Nordling had provided in France. In the meantime the family managed to get the sequestration order accepted and in consequence a delay ensued in the settlement of the estate which was to entail considerable losses.

A sequestration order issued by the family in Britain met with no success, however.

During my stay in San Remo I supervised the dispatch of Nobel's library and laboratory equipment to Bofors. His private possessions, including jewellery and trinkets, were sent to Ludvig Nobel junior for distribution among the family.

Outstanding business transactions, among them the recovery of patent fees in Italy, were entrusted to the firm of bankers Marsaglia & Co. used by the Swedish Consulate. Finally, and in consultation with Emanuel, we presented the Consul himself with Nobel's recently purchased thoroughbred, as a mark of gratitude.

My wife and I then returned to Sweden via France and Germany, stopping for consultation with Nordling and Coulet concerning French inheritance laws.

Owing to the possibility that Nobel's relations might actually start a lawsuit in Paris, we had decided to engage a better-known lawyer than Coulet and hence Nordling had put me in touch with the famous solicitor, senator and former Home Secretary M. Waldeck Rousseau, having already informed him about the circumstances of the will in France and the steps we had taken. During talks at his home in Paris he argued that Bofors was the only alternative to Paris as Nobel's *domicile de fait*. Meanwhile he thought it possible in the event of a lawsuit to establish the validity of the will before a French court—although this appeared to us as over-optimistic.

In Hamburg I was able to interview the German dynamite

companies about the winding up of some of Nobel's interests in South Africa and to settle the question of royalties for ballistite in Germany and Great Britain.

The Struggle Intensifies

The great quarrel over the validity and implementation of the will led in the spring of 1897 to some preliminary skirmishing between the parties concerned. Officially, they took the form of two 'friendly actions' (mock trials) brought against Hjalmar and Ludvig Nobel by Professor Hjalmar Sjögren and his wife, their claim being that the legacies granted to the former two should be declared null and void, so as to safeguard Mrs Sjögren's future inheritance.

The case against Ludvig was heard at Stockholm's City Court, and that against Hjalmar at Karlskoga's District Court. In both instances the defendants contested the authority of the court. It was contended on Hjalmar's behalf that if Alfred Nobel, at the time of his death, was domiciled in Sweden, this must have been in Stockholm, where he was last registered. Against the will of both plaintiffs and defendants, the executors now appeared as 'the intervening party', supporting the claim of Ludvig Nobel that Alfred's legal domicile at his death must, on the contrary, have been Bofors, where he had his household, a permanent staff and comprehensive financial interests. The decision of both courts favoured Karlskoga as the competent tribunal; this meant, of course, that the objection to the 'intervening party' had been overruled. The verdict was the same in the Court of Appeal and the Supreme Court of Justice, where the plaintiffs later lodged an appeal. Before the dismissal of the case by a Royal Ordinance of 6 April 1898 the executors were in addition able to announce that a court in Paris dealing with claims on the estate had declared itself incompetent, on the grounds that Nobel's *domicile de fait* at his death was not Paris, but Bofors.

A campaign against the implementation of Nobel's ideas was, nevertheless, continued in other circles as well. A series of three leading articles was now published in the newspaper *Our Country (Vårt Land)*. These were clearly composed by an eminent lawyer who had been in contact with Professor Sjögren during the lawsuits in Stockholm and Karlskoga. The main argument put forward was, as usual, that the will was unacceptable on purely formal grounds, and since in addition its provisions violated the laws of inheritance and were manifestly unpatriotic, it should also be condemned for moral reasons.

The article went on to praise the cunning strategy whereby the lawsuits in Sweden regarding competency had been 'arranged' for the purpose of spreading the idea that the will was solely an object of dispute among the heirs. This would clearly obviate any criticism likely to arise abroad following a direct attack on the donation, and prevent the executors from dealing with Nobel's property in other countries.

The author also strongly contested the right of the executors with regard to probate. Admitting that they were acting in accordance with established practice, he stated that this was nevertheless incorrect in principle since it was based on the supposition that they represented a beneficiary which did not in fact exist, namely the fund. The executors could in fact therefore act neither for Nobel himself as a deceased person nor for prize-winners who had not yet been designated. Only the heirs were qualified to administer the capital and make such changes in the provisions of the will as they deemed necessary.

These skilful arguments made a considerable impression on the general public, particularly on the prize-giving institutions, and the executors thought it essential to compose a reply. This was penned by Lindhagen, and published in a leading Swedish newspaper in April 1897. Defending the executors' right to implement the will, he claimed that their authority was founded on the task spelled out by Nobel and already established by practice as well as by the courts:

> People speak loudly about the necessity of changing the will in a more patriotic direction, meaning the exclusion of the prize for peace which, it is implied, if distributed by the Norwegian Parliament, may involve some sort of danger for Sweden. Yet surely the only persons who have the right to complain about the dispersal of a fortune are those who have helped to create it! Nobody represents them in this case. It might also be asked how one could reasonably expect a benefactor who, since the age of nine, had spent most of his life abroad and made most of his money there, to do more for Sweden than to set up the administration of the donation and the distribution of the prizes in that country! One might also be of the opinion that the international character of the proposed fund, in itself a lofty conception, would make for a better understanding between the nations.

A second article by Lindhagen ended with a sharp attack on the family, whose members, according to the article in *Our Country*, 'had gone well-equipped into the fight': 'It cannot either be denied,' wrote Lindhagen, 'that they must be pretty thick-skinned if they can go the whole hog in this matter without feeling the worse for wear. But whether they land on the soft cushion of paper money, or on the hard rock of disillusionment, one thing is certain; whatever they gain, it will not be a heroic reputation.'

There was no doubt that Lindhagen's contributions did much to weaken the impression made by the articles in *Our Country*. Nevertheless, I myself felt misgivings over their aggressive tone, which might make a friendly settlement with the relatives even more unlikely. I therefore cabled Lindhagen asking him to consult either me or Lilljequist before continuing his polemics in the press—a request which Lindhagen took as unjustifiable criticism.

He wrote to Lilljequist:

Dear Rudolf,

I have just received Ragnar Sohlman's telegram from Paris. Surely you realize that when you are in the midst of events, and it is essential to steer public opinion in the right direction, there can be no question of simply remaining passive. As for consulting the executors in a hurry, they are usually in different parts of the globe! It is obvious, too, that anyone who sees a situation at close quarters is in a better position to judge what is needed. The articles in *Our Country*, with their special terminology, caused a flutter in the various academies—and the concept of coming to terms with the heirs through relinquishing a part of the prize money is put forward now and again, as well as a mass of other muddled proposals.

Both Ragnar's brother Harald and *Dagens Nyheter* [a Liberal newspaper] rang me up, urging me to reply. I then wrote two articles, aimed at putting matters in their right perspective. This is the only way to counteract plans which take advantage of public ignorance and are usually made with malicious intent. I can assure you that my articles were effective—and things have now quietened down considerably. I hear, too, that the family is feeling a bit crestfallen, now that it has been made clear—perhaps for the first time—that it may have been building castles in the air.

Leman, the lawyer acting for the heirs, now asked me to see him about a possible settlement. He was quite amenable, and told me he knew very little about the case; it seems clear that both the heirs and their advisers are uncertain as to ways and means. Leman declared that the measure for sequestration which had been taken in Paris and also supposedly in Scotland, Germany and Austria, really served no useful purpose, and that he would therefore advise the family to allow the executors the right of disposal over everything that did not directly concern their interests, on condition that nothing be either sold or converted before probate had been granted. He admitted that sequestration had been a deliberately hostile action by the heirs, due partly to their anger with the executors for taking it upon themselves to sell 'family' papers, i.e. shares in the dynamite companies; the relatives also resented the appropriation of Nobel's correspondence, which they considered their property, and the fact that they had been ignored at the sale of his personal possessions in Paris, since this had meant that they were themselves obliged to bid for such articles as they wished to keep.

I replied that the family had no reason to complain. I personally had nothing to do with any sale of securities, but promised that if they could present a serious proposal for a settlement it would be put before the executors. I also suggested that such a proposal should include an offer to withdraw all objections to the decision of the two courts on the matter of competency.

Leman has asked to see me again after further discussion with his clients. He is clearly afraid that people will accuse him of making unnecessary trouble for the executors.

As far as opposition to the will itself is concerned, Leman could only say that an agreement is still far away. For my part, I think it unlikely that Sjögren and Ridderstolpe will withdraw completely, having once entered the fray.

Things seem to be developing satisfactorily in the Academy of Sciences, in spite of efforts by the Caroline Institute to persuade them to their way of thinking. The Institute will claim an annual fee of at least 50,000 crowns for expenses, although nothing is definitely decided yet in any of the academies . . . Meanwhile, Leman has asked me to keep our negotiations confidential, and I agree that it would be a bad thing if the family got to hear of them through other channels . . .

Though somewhat relieved by Lindhagen's official statement that he had 'done nothing other than contribute faithfully, in the discharge of his duties as a Swedish official, to the safeguarding of Nobel's interests', Nordling's fears about his part in the story were only finally put to rest through a decree by the Swedish government of May 1897, instructing the Attorney General to make the legal arrangements necessary to give validity to the will. This was the result of a memo presented to the government in April 1897 by the executors, together with a copy of the will 'for information and possible action'. In connection with a long and formal presentation of the situation as such, the Attorney General explained that the Swedish government wished to co-operate in the carrying out of the will, on the grounds that the case was exceptional and of particular importance, and that every step would be taken to facilitate the implementation of Nobel's intentions.

The attitude of the academies

The government now requested the Swedish Academy, the Academy of Sciences, the Chancellor of Stockholm University and the Caroline Institute to co-operate with the Attorney General in the matter.

Probate was finally granted at Karlskoga's District Court on behalf of the Swedish State and the relevant institutions. At the request of the executors, the Norwegian Parliament was also represented.

The decision of the Attorney General must have been a blow to the plaintiffs as well as a rebuff to the author of the articles in *Our Country*.

In order to effect a radical change in the provisions of the will, all that remained to the family now—apart from the uncertain prospects of a direct appeal in a Swedish court —was the hope of persuading the prize-giving institutions to co-operate in such a change. An opportunity arose while the academies were considering the letter from the executors regarding the said provisions. As we know, there were considerable differences of opinion in the academies concerning the proposed task, and fears voiced that the Foundation would be detrimental to Swedish research and to the prize-giving bodies themselves. Discussion was particularly stormy in the Swedish Academy and the Academy of Sciences, and there were members of the Caroline Institute who wanted a reconstruction of the whole donation, to allow each institution to use its share of Nobel's money at its own discretion—an aim which could only be achieved through co-operation with the family in their bid to have the will declared invalid.

Opposition in the Swedish Academy was headed by B.E. Malmström and President Hans Forssell, who queried the testator's expression 'The Academy in Stockholm', asking whether this really meant the Swedish Academy—a question which could only be decided in a court of law. Malmström argued that the Academy would have to consider whether it could accept the assignment at all; whether its members were capable of appraising the annual literary output of the world, or even of Europe alone; and whether it was prepared to expose itself to all the unpleasantness, pressure and slander which would certainly result from such a task. Even if the answers to these questions were positive, the approval of H.M. the King would have to be sought, since the extra work involved was likely to demand both time and attention; this would inevitably interfere with the main functions of the institution and transform it from a purely Swedish academy into a cosmopolitan tribunal.

For the time being, the Academy ought therefore to take no part in any action concerning Alfred Nobel's will.

Forssell's comments were in the same vein, although even sharper, ending with a demand that the letter from the executors should be ignored.

The chief supporter of a positive attitude towards the donation was the Academy's permanent secretary, C. D. af

Wirsén, whom I had known since childhood and with whom I had often spoken about Nobel's ideas.

After refuting the objections expressed by Malmström and Forssell, he made the following statement:

> If the Academy rejects this charge, the idea of a literary prize as imagined by Nobel will come to nothing, and the foremost contemporary writers on the continent be deprived of the exceptional recognition and privileges he intended for them. A storm of criticism and reproach will undoubtedly follow. In shirking this great responsibility the Academy will be bitterly blamed, and it is probable that our successors in this body of eighteen men will find it strange that we abstained from accepting an influential position in world literature in order to save ourselves trouble.
>
> The charge is said to be incompatible with the duties of the Academy.
>
> Admittedly it is new, and very comprehensive, but being of a literary nature it is certainly not foreign to our work. Without extensive knowledge of the best contemporary literature in other countries, no academy can judge its own—and after all, the prizes in question are intended for those men of outstanding literary merit whose work should be familiar to the Academy.

During the ballot which followed, Wirsén received twelve votes for his proposal, while Malmström and Forssell got only two each. Shortly afterwards their objections were overruled by the Attorney General's official instructions to the Academy to attend probate on behalf of the government.

On my return from abroad in May, Lindhagen advised me to call on some of the more influential members of the prize-giving institutions, among them—apart from Wirsén—President Forssell and Professor Key, the head of the Caroline Institute. Forssell explained that he personally was opposed in principle to the concept of the academies accepting Nobel's charge, but that he knew that he had little support for this view. He now declared that he would accept a decision whereby the academies would shoulder the task on condition that all unsettled problems could be suitably solved. He did not, however, explain how this should be done, stating that any spelling out of these conditions would simply give the heirs reason to complain that they were contrary to the wording of the will.

The other members present expressed their willingness to support the request of the executors. Meanwhile, Professor Key showed me the answer which he intended to present to the board of the Caroline Institute, which made it clear that the Institute would accept the charge if additional explanations and directives be provided.

Soon after, the Academy of Sciences announced that it would accept on the same terms.

Nevertheless, when the matter was discussed in this academy, Forssell succeeded in rejecting the proposal of the relevant committee and persuading the Academy not to take a definite stand until the will had become legally valid. He also urged the rejection of the executors' request for delegates to discuss explanatory additions to the will, his argument being that any such additions might encourage the heirs to start a lawsuit. Owing to the negative attitude of the Academy, the question of the validity of the will was now in a vicious circle—a development which Forssell had obviously foreseen in his determination to have the matter dismissed.

For probate to be granted it was an absolute condition that all the institutions mentioned must accept their charge —otherwise the dispositions concerning them would become null and void. Now that one of them refused even to send a delegate to negotiate with the executors, the possibility of an 'official' agreement was blocked, and deadlock threatened.

The winding up of the estate. Protests by the heirs

I had kept in touch with Emanuel Nobel by letter since our encounters in Paris in the winter and spring of 1897, and he had often suggested that we should meet somewhere for further talks. At this time (April 1897) he was busy negotiating with the German banks concerning a debenture loan of ten million roubles on behalf of the Naphtha Company. Our earlier correspondence had mainly dealt with the financial matters of the estate, especially the company's major liabilities to Nobel himself. I had, of course, assured him that every consideration would be given by the executors to the winding up of these matters

Now Emanuel was worried by a formal intervention from the Swedish Consul General in St Petersburg with regard to Nobel's property in Russia.

On 3 April 1897 he wrote to me from St Petersburg:

Dear Ragnar,

Without referring to the questions raised in your welcome letter from San Remo, I must return to what I said about the Swedish–Norwegian Consul General here, a man called Damberg. The situation is particularly irritating since I remember telling you not to bother too much about the funds still in Petersburg.

Damberg, who is a fine old fellow and a great friend of mine, would rather not be mixed up in the affair, but he is afraid of criticism from Stockholm for not having paid close enough attention to the question of the Nobel capital. Apparently it is his duty to make some sort of 'declaration' to the Russian authorities, who will then call on me, in the company of the Swedish Consul, and I shall be obliged to hand over the whole fortune to them. I had not taken this seriously, but now it seems that Reuterskjöld [the Minister] who is a nervous fellow, is pushing the matter with excessive officiousness . . .

I think the best idea would be for you to write to Damberg, or to me, saying that nothing should be done until you come over yourself, or until the will has been recognized in Sweden . . .

Affectionately, E. Nobel

My proposed visit to St Petersburg was postponed until the following June when I invited a former secretary from the Russian Nobel Company to accompany me as an interpreter. Arriving on board the steamer, I was amazed to find Hjalmar Nobel and the Ridderstolpes there. Hjalmar's first words on meeting me were: 'So you are off to Petersburg—well, so are we!'

The company thus forced on us in St Petersburg was not especially agreeable either to Emanuel or me. The cousins did not get on too well, and were suspicious of each other. Hjalmar insisted on being present at every discussion I had with Emanuel, and in such a strained atmosphere there was little chance of mentioning the will. A pleasant interlude for me was a visit to the Nobels' beautiful home, where I was warmly received by the family—Emanuel and Edla Nobel, and their young relatives. Relations with the Consul General improved, and the necessary steps were taken for an inventory of the property of the estate in Russia. The combined value of these assets, included in the final inventory in Sweden, amounted to 5,232,773:45 Swedish crowns.

The Academy of Sciences' refusal to take any immediate action made it even more essential for the executors to come to terms with the heirs. Both Lindhagen and I favoured a friendly settlement, but Lilljequist was extremely sceptical of reaching any acceptable solution, and also questioned our authority, particularly in the interests of future prize-winners, to involve the estate—in other words, the future fund—in any financial sacrifices or concessions. Paradoxically, his attitude in this respect helped to strengthen the position of the executors at the final settlement.

To gain a better understanding of the immediate situation and the prospects in case of a lawsuit abroad, we decided in the beginning of June 1897 to arrange a meeting with our foreign advisers.

Coulet from Paris, Lindhagen from Stockholm, Warren from Glasgow and Westphal from Hamburg now met up with us in Stockholm. Coulet and Warren considered the situation both in France and Great Britain favourable, since the executors did not need to fear any legal proceedings there. Westphal, on the other hand, who was obviously influenced by Max Philipp, urged an immediate compromise with the heirs. This conference with four prominent lawyers, each one typical of his own nationality, was both interesting and amusing. The most talkative turned out to be Coulet, whose *discours* in the grand manner, invoking every reason and prospect for the carrying out of Nobel's intentions, was a masterpiece of oratory. It was warmly applauded, and even Warren remarked, with dry Scottish humour: 'The little chap speaks quite nicely'.

There were various proposals for a friendly settlement with the heirs, but they led nowhere. During a talk I had with Leman, he suggested that the executors should relax their insistence as to the validity of the will, so that the property of the estate would automatically go to the natural heirs. In return, the latter were to agree among themselves to assign a large part of the inheritance for the fulfilment of Nobel's wishes. This would benefit them, as it meant a considerable reduction of the inheritance tax, both in Sweden and abroad (which finally amounted to more than three million crowns). Being young and inexperienced, I answered rather abruptly that an agreement on these grounds would be generally looked upon as a *pactum turpe*, and this offended Leman so much that he refused to have anything more to do with the executors.

Meeting with Ludvig Nobel shortly afterwards, I confronted him and asked what he and the family really meant by their opposition to their uncle's will, and what they thought were their rights in it. He replied that the relatives should have been allotted 'the family papers', in other words all shares in the dynamite companies in the different countries, the shares in the Naphtha Company and in Bofors, and finally the house in Paris. Since these assets totalled about a third of the estate, I refused, again perhaps too abruptly, to carry on the discussion. As a result my long friendship with Ludvig was disrupted, and only re-established during the last years of his life.

The size of Nobel's estate and the complications involved by its dispersal in so many countries caused great delay in the winding up of his affairs, and the executors had to apply repeatedly to the District Court at Karlskoga for postponement of the final Deed of Inventory. In this connection, my mind goes back to two serious interruptions to my work during this time.

As we know, Nobel had a passion for thoroughbreds, and at Björkborn he had kept three magnificent Orloff stallions. These animals, however, were bred on the plains, and found the hilly countryside in Värmland an unpleasant surprise. Trotting downhill, they had difficulty in holding back, since it strained their graceful forelegs, and this meant that they often had to be kept indoors for treatment. One day, when my wife and I happened to be driving to the station, one of them kicked over the traces, and bolted; the dogcart overturned as we dashed round a sharp corner, and fell on top of me, causing two broken ribs and injuring my back. My wife and the groom were unhurt, but I had to spend several weeks in bed, which upset all my plans and prevented me from going to San Remo for the sale of Nobel's villa there.

The second interruption, which occurred about a year later, was due to my becoming liable for military service. On leaving Sweden for America in 1890, I had been too young for this, and did not give it a thought until years later. When, after some time at Bofors, I mentioned to Nobel that I would probably have to join up, he retorted that our experiments with gunpowder and firearms were of greater importance to the army than 'your drilling as a recruit'. After his death in 1896 my time was so taken up with journeys and the duties of an executor that I forgot all about the matter. But in the spring of 1898, army regulations caught up with me, and I could not put it off any longer. An application to the authorities for exemption was turned down and, at the beginning of May, I had to present myself at the headquarters of the infantry regiment in question for a preliminary training period of seventy days.

Since the current business of the estate could not be held up, and Lilljequist was not able to take over, I had to install myself and my papers in a private building inside the camp, where my accountant and I rented a room each where, after the day's work, we could attend to correspondence. An elderly vice-corporal was detailed as bicycle orderly to post my letters at the neighbouring station of Hallsberg. The telephone presented a serious difficulty at first—the only one available being in the officers' mess—and incessant calls from banks and stockbrokers to Recruit 114 Sohlman caused great irritation, especially during exercises. The problem was finally solved through a written agreement between 'the non-commissioned officers of the Lifeguard Regiment, the local tradesman Olsson, and Recruit 114 Sohlman'. According to this, the said officers, who had been permitted by their seniors to extend a cable to their mess by means of some already existing telephone poles, were willing to let Recruit Sohlman install an apparatus in his temporary quarters. Olsson would pay a ground fee for the use of the phone during the day, and when Recruit Sohlman was duly demobbed, it would be handed over to the non-commissioned officers' mess room.

This convoluted arrangement gives a good picture of the facilities for telephoning in Sweden at the time.

Urgent business in Stockholm obliged me to apply for leave three times. The last occasion concerned a visit to Kristiania (Oslo). Saluting my company commander, I respectfully requested three days' leave for this purpose. 'What's that?' he barked, 'what the hell has Recruit Sohlman got to do in Kristiania?' My answer that I had been summoned to negotiate with delegates from the Norwegian Parliament caused no little surprise. (The crisis between Norway and Sweden at that period, with both countries preparing for war, was tense.) However, when the matter came up before the commanding officer, I was given the necessary leave, and told to go to Hallsberg, change out of uniform and 'disappear'—a hint as good as any to desert . . .

Work on the winding up of the estate had continued in the meantime, so that the heirs could be summoned to the inventory proceedings on 30 October 1897. Only a few of them actually attended. The District Judge represented Robert Nobel's widow, Mrs Pauline Nobel, Hjalmar and Ludvig Nobel and Count Ridderstolpe with his wife. Carl Lindhagen acted as Administrator, and Lilljequist and I formally handed over the estate.

The representative of the plaintiffs then delivered a written protest against the inventory proceedings, using the same arguments as before—namely, the uncertainty about Nobel's true domicile at the time of his death and hence the lack of a competent court to handle the case, and finally the effrontery of the executors in not inviting the family to the inventory proceedings in France.

The protest was not upheld, however, and after the conclusion of the proceedings, the documentation was duly delivered to Karlskoga's District Court, on 9 November 1897, eleven months after Nobel's death.

In accordance with an official form presented to us by Lindhagen, we placed a final summary of Nobel's assets and liabilities at his death under two main headings:

A. Assets not subject to inheritance tax abroad, valued at	18,123,043:92 crowns
B. Assets taxable abroad, valued at	15,110,748:78 crowns
The liabilities of the estate were assessed at	1,646,589:92 crowns
The net assets thus amounted to	31,587,202:28 crowns

According to the estate inventory, the disposition of the assets, i.e. their placement in the different countries after Nobel's death, were as follows:

Sweden	5,796,140:—
Norway	94,472:28
Germany	6,152,250:95
Austria	228,754:20
France	7,280,817:23
Scotland	3,913,938:67
England	3,904,235:32
Italy	630,410:10
Russia	5,232,773:45

The executors claimed that Swedish inheritance tax, after deduction of the liabilities, should only be imposed on the assets listed under A. This claim was quashed by the District Court, and the executors ordered to pay inheritance tax on the total assets, whether or not part of these had previously been taxed abroad. An appeal to the Supreme Court only resulted in a confirmation of the District Court's ruling, even though the President of the Court dissented, supporting the claim of the executors. The ruling of the District court was finally upheld by a Royal Ordinance of 10 April 1899. Thus, the inheritance tax to be paid in Sweden amounted to 1,843,692:25 crowns, while the tax payable abroad came to 1,235,949:96 crowns, the total tax amounting to 3,169,642:21 crowns.

The securities deposited with the Disconto-Gesellschaft in Berlin include the following Russian assets:
Russian government bonds for 511,589:25 crowns
Shares in the Nobel Brothers Naphtha Co. for
1,691,000: crowns

After the conclusion of the Deed of Inventory, there was still work to be done with regard to the sale of property and the finalization of some of Nobel's projects.

Some delay was now caused by the sequestration measures

taken in France and Germany by Hjalmar Nobel and other members of the family. They had also protested against the granting of probate in England, although for reasons unknown, not in Scotland where papers, valued at much the same figure as those Nobel had placed in England, had been deposited.

Probate was finally granted in Scotland in August and in England four months later. These formalities, however, only applied to securities held in the two countries at the time of Nobel's death, and did not affect the documents transferred from Paris by the executors and deposited in London in their name.

The sale of the latter papers, particularly the speculative ones, i.e. gold-mine shares, Brazilian and Argentinian government bonds, Spanish and Portuguese railway stock, and shares in the *Banque Ottomane* and other banks, was completed during 1897 and the proceeds transferred to banks in Sweden.

The above-mentioned sequestration measures caused the estate much unnecessary expenditure; we had to pay the running costs of the house in Avenue Malakoff, and were prevented from exploiting favourable openings for the sale of securities. Leman was well aware of the situation, and now suggested to Lindhagen the possibility of a gentleman's agreement whereby the heirs would annul the sequestration orders abroad in return for a promise by the executors to postpone the sale of the so-called 'family papers', i.e. the shares in the Nobel Dynamite Trust Co., the Naphtha Co., and Bofors, pending a court decision on the validity of the will.

I myself negotiated for several months with Hjalmar on the matter, though with little success, since he suspected that any concession by the heirs would prejudice their interests. Our French adviser, Coulet, suggested that we should take steps to annul the sequestration in Paris, but both Lindhagen and Nordling opposed this on the ground that it would look like an open declaration of war.

As a result of the delays in the sale of the Dynamite Trust shares the estate suffered a considerable loss—having been fixed at a high figure in the summer of 1897 they lost ground in the autumn of the following year, obliging us to reduce their total value by 182,500 crowns in the final inventory.

The End of the Struggle and the Birth of the Nobel Foundation

Enterprises supported by the estate after Nobel's death

Nobel had backed a number of other people's inventions, and these now caused the executors considerable worry. We had already decided that all activities in the Björkborn laboratory, including his own experiments, should be carried on for at least a year after his death. One difficulty was the evaluation of these projects, and thus of establishing the amount of money which the estate could expect from Nobel's share in them. In several cases the inventors concerned persuaded us to take an over-optimistic view of their work.

Examples of such projects were:

A. A contract with R. W. Strehlenert, dated 15 July 1896, regarding a method for manufacturing artificial silk—which was later dropped as impracticable.

B. A contract with W. Unge, dated 15 July 1896, concerning the invention of war rockets. Krupp was interested for a time and carried out some experiments, which led to no immediate result but were further developed in later years (see page 30).

C. A contract with B. and F. Ljungström regarding a bicycle with variable gears.

D. A contract with B. and F. Ljungström, dated 28 July 1896, regarding one third of the profits on a steam boiler and condenser.

(In addition, the inventory included a commitment of quite a different kind, namely an agreement to participate in 'The South African Explosives Agency'—in view of existing conditions in South Africa an extremely risky obligation from which, after considerable negotiations in Hamburg, we managed to free the estate without loss, and even with a certain profit.)

The bicycle with variable gear change, on the other hand, gave rise to much trouble and lengthy negotiations before the comprehensive investments made in it by Nobel could finally be liquidated. This elegant invention, which was considered a great novelty at the time, was highly recommended by the famous British inventor, Sir Hiram Maxim, and awarded a gold medal at the Stockholm Exhibition in 1897. Later, several of Nobel's wealthy friends in England and Scotland, including Lord Ribblesdale and Sir Charles Tennant who were members of the Dynamite Trust, helped to form a company with a share capital of £200,000, named the New Cycle Co., to promote the so-called Svea Bicycle. Nobel himself bought shares for £40,000 in the enterprise, and his portfolio at his death, which had been extended by the shares he had received as payment for the invention, amounted to £75,074.

The company set up a factory with a capacity of 100 bicycles a week at Northfleet on the south bank of the Thames. But bad management prevented the venture from prospering, and in view of the large financial interests involved, Lilljequist and I decided to go over to London for a few weeks to help our Scottish lawyer, Warren, to find a new manager, restructure the business and inject more capital into it. We were, of course, unable at this point to take any direct action, as probate had not yet been granted in England and the estate could therefore not be officially represented at company meetings.

In spite of our efforts, the company foundered, chiefly, I think, because the invention itself did not come up to expectation. The vertical pedalling movement on which it was based proved to be more tiring in the long run than the conventional method.

This venture resulted in an estimated loss to the estate of

567,329 crowns, a conservative figure considering the work that Nobel had put into it.

In addition to this account of the settlements of Nobel's financial interests as a patron of other inventions, I will now give a brief survey of the experimental work which he himself was pursuing at San Remo and Björkborn at the time of his death.

The main fields of activity consisted in:

Artificial caoutchouc or rather attempts to develop a substitute for natural rubber and for leather based on nitrocellulose plus suitable gelatinizing agents.

Progressive powder, i.e. a smokeless powder in which the separate powder grains or powder pipes consisted of layers having differing and gradually increasing combustion speeds.

Ballistite (or Nobel powder) with lower combustion temperatures and reduced corrosion.

Firing trials with Nobel powder to determine the pressure generated at varying charge densities.

Production of nitrocellulose suitable for the preparation of rayon, with associated trials.

Projectile sealing, i.e. attempts to reduce gun-tube corrosion by using copper girdles, which on firing expanded from the effect of the gas pressures produced.

A rocket camera or photographic telemeter, an attempt to achieve photographic mapping of land areas by launching a parachute rocket or captive balloon carrying a camera, the exposure being triggered by fuse or by electrical means.

Some of these ideas have been adopted in later years—as, for instance, photographic mapping of land areas, an important factor in connection with the development of aerial research.

Work in the first four of these fields continued at Björkborn for several years, firstly under the management of the estate and later under the supervision of the newly-created company, Bofors Ballistite. New types of gunpowder were developed, and from 1899 onwards used in Sweden and Norway for heavy artillery in the navy, to replace imports from France and England. The firing trials were significant for the improvement of the method of internal ballistic calculations known as 'the Bofors method'.

The expansion in productivity at Bofors Ballistite as a result of all this considerably increased the royalties accruing to the estate, and brought many thousands of crowns both to this and, later, to the Nobel Foundation.

In the spring of 1897 negotiations were opened with the Nitroglycerine Company for the transfer to Bofors of the manufacture of ballistite from the company's explosives factory at Vinterviken (south of Stockholm), a plan I had already discussed with Nobel during his lifetime.

The Nitroglycerine Company had acquired the sole right to make smokeless gunpowder in Sweden according to Nobel's patent (the royalties amounted to one crown per kilogram sold) but lacked the technical know-how to exploit the invention. Other reasons for the proposal were the lack of artillery ranges and, above all, shooting expertise at Vinterviken, whereas Bofors was well equipped in both these respects. The negotiations were finally concluded in 1898 with the formation of the previously-mentioned Bofors Ballistite Co. in which the main subscribers were Bofors and the Nitroglycerine Co. with the estate as third party. It was a modest beginning, with a share capital of only 180,000 crowns, of which the Bofors–Gullspång Co. and the Nitroglycerine Co. subscribed 80,000 each and the Nobel estate 18,000, the rest being provided by private individuals. In addition, the principal shareholders made certain concessions without special remuneration—Bofors by providing land for the future powdermill and the use of the company's shooting ranges, Nitroglycerine by surrendering the manufacturing rights for ballistite, and the estate by contributing some of Nobel's patents. To begin with the new company was permitted to use the laboratory at Björkborn (which it later took over) free of charge. I myself was appointed managing director, with a salary of 3,000 crowns a year and a small commission on the company's future profits. I was also given free lodging in Nobel's manor house at Björkborn, where I was to spend the next twenty-five years.

The company is now an integral part of the main Bofors concern.

The winding up of Alfred Nobel's interests in Bofors

The sale in 1898 of Nobel's shareholdings in the Bofors–Gullspång Co. was to have far-reaching consequences.

His contribution to the development of the Bofors iron-works when he was sole owner had been of major importance to the business. His death was thus a harsh blow, and it was clearly essential both for the company and the nation that the future ownership be placed in secure hands. At first I toyed with the idea of retaining the majority of the shares for the Foundation which, after all, was to be the chief beneficiary of the will. In this context we considered a proposal which Nobel himself had discussed during the last years of his life, consisting of a merger between Bofors and some foreign munitions manufacturer.

An opening presented itself in 1897, when a representative of the large Palmer Shipbuilding Co. in Britain contacted me with regard to a possible merger between his company and a well-known British manufacturer of armour plating, Brawa & Co., and Bofors. The intermediary was a Swedish businessman living in England named Peter Pehrson, and in 1897 we visited together the Palmer Shipyard at Jarrow-on-Tyne to discuss the possibility of the two British companies and Bofors forming a joint enterprise, in which each would retain their independence but combine as shareholders. The new company was to provide warships with full artillery equipment, etc.

After returning to Sweden, I discussed the proposal with the new manager of Bofors, Commander Dyrssen, (who later became Marine Minister). He seemed interested at first, but after conferring with Prime Minister Boström, who opposed the idea, he advised against continued negotiations with the British and the matter was dropped.

The idea of retaining the Bofors shares for the Foundation turned out to be impracticable, since it was more than likely to have an adverse effect on the development of the company. It must be added that the manufacture of munitions could in any case scarcely be said to comply with the aims of the Foundation. Thus it became necessary to find a Swedish buyer or rather to form a Swedish consortium to take over the business, a complicated matter in view of the size of the enterprise.

When Nobel died, the share capital in the Bofors–Gullspång Co. consisted of 4,000 preference and 6,000 ordinary shares at a par value of 500 crowns each, totalling five million crowns. He had held all the shares himself with the exception of 400 ordinary shares which had been registered in the name of Hjalmar Nobel.

To reduce the size of a sale and thus facilitate the creation of a Swedish syndicate to take over the ordinary shares, the executors proposed an agreement between the board of the Bofors Company and the Nobel estate for the conversion of the preference shares into three and a half per cent bonds, which would remain as part of the estate for the portfolio of the future Foundation. In accordance with this plan it was decided in January 1898 to issue a bond loan of two million crowns in payment of these.

In September the same year the sale of the ordinary shares was finally concluded, as described by the historian Steckzèn:

> The British firm, Palmer's Shipbuilding Co., which wished to acquire Bofors, was represented by the Swedish–English financier, William Olsson, who did every-

thing possible to force a decision. The executors were reluctant to reject his bid, but nevertheless reserved the right to negotiate with others in the hope of establishing a purely Swedish consortium. From the company's viewpoint, the best bet was the Kjellberg family, which had strong personal ties with Bofors . . . At the end of August Sohlman suggested to Jonas Kjellberg that the latter should set up a consortium to take over the ordinary shares. Kjellberg, however, turned down the idea, stating that he and his family did not have adequate means for a purchase of this magnitude.

A few days later, Kjellberg went on other business to Gothenburg, where he encountered a bank manager named Theodor Mannheimer. The latter inquired, in passing, if the Kjellberg family were willing to buy the Bofors shares, and when the position was explained to him, offered to lend Kjellberg the necessary funds against the security of the shares in question, adding that he himself could place any part of the capital stock which the Kjellbergs were unable to buy. As a result, Kjellberg immediately entered into fresh negotiations with the executors.

The final settlement took place in dramatic circumstances. Jonas Kjellberg tried to beat down the price of the ordinary shares as far as possible, and to that purpose prolonged the discussion, while William Olsson sought to force an acceptance of the executors' conditions in the interests of the British syndicate. Lilljequist and [Sohlman] therefore decided to bring matters to a head by giving Kjellberg the option of buying, with a deadline at 9 a.m. on 15 September. If he did not reply by this time, Olsson's bid would be accepted.

The situation remained fluid right up to the last day. At 8 a.m. on 15 September Sohlman, who was in Bofors, finally received an emergency call from Kjellberg in Gothenburg, accepting the offer, while Mannheimer simultaneously confirmed that the money would be paid over by the Scandinavian Bank. At 8.15 a telegram arrived from Kjellberg, tersely stating: 'I accept', thus spelling the end of a singularly exhausting transaction.

The agreement was confirmed by a comprehensive contract on 10 October 1898 whereby the Kjellberg consortium bought the ordinary shares for 1.5 million crowns, a sum corresponding to 50 per cent of their nominal value. In a further agreement of particular importance to the future of Bofors, the estate sold its own shares in the newly-created Bofors Ballistite Co. to the Bofors–Gullspång Co., thus assuring the latter the share majority in the former. At this point Kjellberg gave a verbal assurance on behalf of the majority of the shareholders—a kind of gentleman's agreement—not to allow the share majority to pass into foreign hands. This promise was strictly adhered to, although favourable offers were later made on several occasions by foreign buyers.

The settlement between the estate and the Kjellberg consortium meant that Bofors—Gullspång and Bofors Ballistite were retained as wholly Swedish companies. This was of great significance, as the two companies were later to provide a useful and necessary counterbalance to the interests of the arms industries in Britain, France and Germany.

It should be added that in financial terms the bid made by Olsson was somewhat more favourable than that which Kjellberg obtained in his time-limited option. The former offer did not, however, include any reservations concerning the transfer of shares to foreigners.

The final price ensured full compensation for the money which Nobel himself had invested in Bofors, and was duly added to the estate.

The end of the struggle and the birth of the Nobel Foundation

After various trips to Hamburg and London to confer with lawyers, I was requested to travel to St Petersburg in December to see Emanuel Nobel. There we discussed the will in general and the possibility of liquidating Nobel's Russian interests, and also touched on the transfer of shares in the Naphtha Co. to Emanuel on behalf of the family.

At the beginning of 1898, we managed to reach an agreement with Nobel's Swedish relatives which enabled us to distribute certain legacies to private individuals named in the will. There had been no real objection to this earlier, although the validity of the bequests had been formally questioned by the family on account of the unsolved problem of Nobel's domicile. Some of the legatees were hard up, and had begged for their inheritance, but as long as the will was not legally recognized the executors had no right to pay out any money. Repeated efforts to make the family give us permission to do so had been fruitless, and it was not until January 1898, after I had sent Hjalmar a draft proposal for an agreement containing every conceivable reservation, that the family gave its sanction to these payments.

In an effort to end the deadlock resulting from the refusal by the Academy of Sciences to appoint delegates to discuss the prize-giving, it was now decided to ask the two delegates appointed by the Swedish Academy and the Caroline Institute's teaching staff to meet Lindhagen and the executors for preliminary talks.

Others who took part in the discussions were two voluntary representatives from the Academy of Sciences who had been members of a committee during earlier debates within the Academy on the same problem.

Following six lengthy meetings in January and February 1897 to discuss the statutes of the future Foundation, I and two other participators proposed the establishment of a scientific institution to be named after Nobel, whose task it would be to assist the prize-giving bodies in the choice of candidates.

Emanuel Nobel was invited to attend the last three meetings as a representative of the branch of the family resident in Russia, and it was on one of these occasions that he made the statement which was to be decisive for the recognition of his uncle's will and consequently for the establishment of the Foundation:

> Mr Nobel stated that he desired to respect the intentions and wishes expressed in the will of his deceased uncle. He would therefore not dispute anything contained in the will. In order, however, to give effect to the lofty intentions of the testator, alterations and supplementary provisions were undoubtedly required, and these could not be introduced without the consent of all the heirs . . .
>
> The other persons present stated that they regarded it as a matter of course that Nobel's heirs should be asked their opinion regarding any such scheme. Mr Sohlman added on behalf of the executors that they had always desired to act in agreement with all the beneficiaries in the execution of their office, and that in view of Mr Nobel's attitude to his uncle's will they would in the future, as they had in the past, regard it as their right and their duty to have recourse to his advice and assistance in dealing with the inheritance, and arriving at a solution of the problems connected with the will.

On 27 March 1898 the executors sent a copy of the minutes to the Swedish prize-giving bodies, requesting them to give their views of the principles which had been agreed. Referring to the fact that Emanuel Nobel's statement had changed the situation, and that the implementation of the will was now assured, at least as far as part of the estate was concerned, the executors repeated their plea that the Academy of Sciences should send delegates to help in working out the statutes of the Foundation and to assist the executors in the case of any lawsuits brought by the heirs.

On 11 May the Academy finally agreed to co-operate.

Meanwhile, the heirs in Sweden had taken action in a Swedish court to have the will annulled, citing not only the executors, the three prize-giving bodies and the Norwegian Parliament, but also the Swedish government as representing the community. The writ, dated 1 February 1898, was submitted to the Karlskoga Court and signed by Robert Nobel's widow and children, by Professor Hjalmar Sjögren on behalf of his wife Anna, and by Mr Åke Sjögren on behalf of his stepdaughters, the children of Emanuel Nobel's deceased brother Carl.

The main point in the action was to ensure that once payment of the legacies to private individuals had been made, the residue of the estate—in the absence of any specific indication in the will—should be handed over to the heirs. The latter further reserved the right to query the final inventory and, in the case of the prize-giving bodies accepting the task of distributing the interest from the capital in the form of prizes, demanded full control over the said capital. An assurance was meanwhile given in which the heirs declared that they would do everything in their power to carry out Nobel's main intentions.

On account of the long respite granted to the Norwegian Parliament as a foreign party, the case was postponed until the autumn, and at a hearing on 29 September it was finally declared that an out-of-court settlement had in fact been reached, as detailed in two documents of 29 May and 5 June. The case was then dismissed.

Emanuel Nobel, who had played such a prominent part in bringing about the implementation of the will, had clearly been very awkwardly placed in the whole affair, since he was acting as guardian to his siblings and was thus responsible for their interests as well. One of them later told me that he had called them to a meeting at which he explained the situation in detail and pleaded for their support for any agreement he might have to enter into in order to protect the family name and their common interests—and that his brothers and sisters had agreed to this, during a highly emotional scene.

His own financial interest in the case was restricted to ensuring for the family the Nobel shares in the Naphtha Co. and thus his own control of it.

A preliminary agreement was now concluded for the sale at par to Emanuel Nobel, on behalf of Ludvig Nobel's heirs resident in Russia, of all Alfred Nobel's shares, including so-called *pajer*, in the Nobel Brothers Naphtha Co. The total sum amounted to two million roubles, at the current rate of exchange 3,840,000 Swedish crowns.

Pressure on Emanuel to dispute the will was exerted not only by his Swedish relatives but also by influential circles in Stockholm. One day in February 1898, he was even summoned to an audience with King Oscar, and later told me that the king had tried to persuade him to get the terms of the will changed, especially the clause about the peace prize.* This, the king said, could only cause trouble, adding: 'Your uncle was talked into this by fanatics, womenfolk mostly'.

E. Nobel: 'Your Majesty may be thinking of Moltke's words: "*Der ewige Friede, das ist ein Traum, und auch ein schöner*" —eternal peace is a dream, but a beautiful dream.'

The king: 'Was that what he said—a dream, but a beautiful one . . .' (he repeated the words twice).

King Oscar went on to point out that the will was vague and impossible to implement, which, he suggested, was the reason why the prize-giving bodies themselves were very hesitant, and why the Academy of Sciences refused to accept the provisions until alterations were made. Emanuel replied that he had recently been negotiating with representatives of all these bodies and that they had now reached a basic agreement on the proposed statutes.

The king: 'Well, I suppose one cannot stop people from getting together and talking, but responsible decisions are simply not arrived at in that manner; in any event it is your

* The king's attitude was clearly influenced by his fears of trouble with Norway in view of the Union Crisis.

duty to your family to make sure that their interests are not jeopardized by your uncle's nonsensical ideas.'

E. Nobel: 'Your Majesty— I will not expose my family to the risk of reproaches in future for having appropriated funds which rightfully belonged to deserving scientists.'

The conversation was not followed up, but Nobel's Russian lawyer was appalled when he heard of it, and urged Emanuel to leave for St Petersburg at once to avoid arrest. Such replies to a king must surely be taken as lese-majesty!

It should be added, however, that King Oscar, once the Foundation had been established, showed great appreciation of its aims, and always personally handed over the prizes at the annual ceremony in the City Hall.

Emanuel Nobel's standpoint caused a definite change of heart among other members of the family, and thus made way for a final settlement of the vexed question of the will. Following a compromise solution put forward by the executors, a first agreement was signed on 29 May 1898 by Professor Hjalmar Sjögren, acting for his wife, and Mr Åke Sjögren on behalf of his stepdaughters, who together represented a certain percentage of the inheritors, whereby each would receive the sum of 100,000 crowns, i.e. one twelfth of the annual income from the estate, in return for their relinquishment of all future claims on the said estate.

What remained now was to come to terms with Robert Nobel's family, as representing fifty per cent of the inheritors.

After lengthy negotiations, an agreement was finally reached which secured payment to the latter of 1,200,000, corresponding to the value of one and a half year's income from the capital of the estate.

This agreement was signed before the Public Notary in Stockholm on 5 June 1898 between on the one hand Hjalmar, Ludvig and Pauline Nobel and Count Ridderstolpe, acting for his wife, and on the other Ragnar Sohlman and Rudolf Lilljequist, whereby the above-mentioned relatives undertook on behalf of themselves and their descendants to recognize the will and to renounce all future claims on the estate.

The agreement of 5 June contained the following stipulations, which were later included in the statutes of the Nobel Foundation:

> That the Articles of the Foundation dealing with the conditions governing the award of the prizes provided for under the will should be drawn up in agreement with a representative appointed by the family of Robert Nobel, and that they should be submitted to His Majesty's Government for approval.
>
> That the following main principles should be strictly adhered to: namely that each of the annual prizes provided for under the will should be awarded at least once in every period of five years, to commence with and include the year subsequent to that in which the Nobel Foundation should come into force; and that in no circumstances should the amount of such a prize, when awarded, be less than 60 per cent of the total amount of accumulated interest available for distribution, and that the amount should never be divided into more than three prizes.

The so-called 'moral' stipulations contained in the final documents were aimed at removing any impression of purely financial bargaining, and there was also a general condition that the agreement should be approved by Royal Ordinance and the different prize-giving institutions.

The compromise was finally accepted by the Academies in the first two weeks of June, and duly endorsed by the Swedish government on 9 September 1898.

The Norwegian Parliament had agreed on 26 April 1897 to accept the task of administering the peace prize, and had given a committee of three a directive to negotiate the details with the Swedish executors; a committee of five was set up the same year to deal with the prize-giving itself. A third committee, headed by President Ullman, met the executors in July 1898 to discuss the settlement regarding Nobel's will, and on 4 July this was approved by the Norwegian Parliament.

Through a Royal Ordinance of 29 June 1900 the statutes of the Nobel Foundation were established, as well as the special regulations governing the Swedish prize-giving committees.

The long struggle over Nobel's will was now at an end. In the light of later experience, the final outcome must be deemed satisfactory.

Certainly the predictions about the dangers and difficulties that would result from the implementation of Nobel's intentions have proved wholly groundless.

The Nobel Foundation is a great asset to our country, and the distribution of the prizes an honour which has extended knowledge of and respect for Sweden and Swedish culture throughout the world.

I myself shall always treasure the privilege of having participated from the outset in the creation and development of the Nobel Foundation.

May 1948 *Ragnar Sohlman*

Appendix
ALFRED NOBEL'S WILL

I, the undersigned, Alfred Bernhard Nobel, do hereby, after mature deliberation, declare the following to be my last Will and Testament with respect to such property as may be left by me at the time of my death:

To my nephews, Hjalmar and Ludvig Nobel, the sons of my brother Robert Nobel, I bequeath the sum of Two Hundred Thousand Crowns each;

To my nephew Emanuel Nobel, the sum of Three Hundred Thousand, and to my niece Mina Nobel, One Hundred Thousand Crowns;

To my brother Robert Nobel's daughters, Ingeborg and Tyra, the sum of One Hundred Thousand Crowns each;

Miss Olga Boettger, at present staying with Mrs Brand, 10 Rue St Florentin, Paris, will receive One Hundred Thousand Francs;

Mrs Sofie Kapy von Kapivar, whose address is known to the Anglo-Oesterreichische Bank in Vienna, is hereby entitled to an annuity of 6000 Florins Ö.W. which is paid to her by the said Bank, and to this end I have deposited in this Bank the amount of 150,000 Fl. in Hungarian State Bonds;

Mr Alarik Liedbeck, presently living at 26 Sturegatan, Stockholm, will receive One Hundred Thousand Crowns;

Miss Elise Antun, presently living at 32 Rue de Lubeck, Paris, is entitled to an annuity of Two Thousand Five Hundred Francs. In addition, Forty Eight Thousand Francs owned by her are at present in my custody, and shall be refunded;

Mr Alfred Hammond, Waterford, Texas, U.S.A. will receive Ten Thousand Dollars;

The Misses Emy and Marie Winkelmann, Potsdamerstrasse, 51, Berlin, will receive Fifty Thousand Marks each;

Mrs Gaucher, 2 bis Boulevard du Viaduc, Nimes, France will receive One Hundred Thousand Francs;

My servants, Auguste Oswald and his wife Alphonse Tournand, employed in my laboratory at San Remo, will each receive an annuity of One Thousand Francs;

My former servant, Joseph Girardot, 5, Place St. Laurent, Châlons sur Saône, is entitled to an annuity of Five Hundred Francs, and my former gardener, Jean Lecof, at present with Mrs Desoutter, receveur Curaliste, Mesnil, Aubry pour Ecouen, S. & O., France, will receive an annuity of Three Hundred Francs;

Mr Georges Fehrenbach, 2, Rue Compiègne, Paris, is entitled to an annual pension of Five Thousand Francs from January 1, 1896 to January 1, 1899, when the said pension shall discontinue;

A sum of Twenty Thousand Crowns each, which has been placed in my custody, is the property of my brother's children, Hjalmar, Ludvig, Ingeborg and Tyra, and shall be repaid to them.

The whole of my remaining realizable estate shall be dealt with in the following way: the capital, invested in safe securities by my executors, shall constitute a fund, the interest on which shall be annually distributed in the form of prizes to those who, during the preceding year, shall have conferred the greatest benefit on mankind. The said interest shall be divided into five equal parts, which shall be apportioned as follows: one part to the person who shall have made the most important discovery or invention within the field of physics; one part to the person who shall have made the most important chemical discovery or improvement; one part to the person who shall have made the most important discovery within the domain of physiology or medicine; one part to the person who shall have produced in the field of literature the most outstanding work of an idealistic tendency; and one part to the person who shall have done the most or the best work for fraternity between nations, for the abolition or reduction of standing armies and for the holding and promotion of peace congresses. The prizes for physics and chemistry shall be awarded by the Swedish Academy of Sciences; that for physiological or medical work by the

Caroline Institute in Stockholm; that for literature by the Academy in Stockholm, and that for champions of peace by a committee of five persons to be elected by the Norwegian Storting. It is my express wish that in awarding the prizes no consideration whatever shall be given to the nationality of the candidates, but that the most worthy shall receive the prize, whether he be a Scandinavian or not.

As Executors of my testamentary dispositions, I hereby appoint Mr Ragnar Sohlman, resident at Bofors, Värmland, and Mr Rudolf Lilljequist, 31 Malmskilnadsgatan, Stockholm, and at Bengtsfors near Uddevalla. To compensate for their pains and attention, I grant to Mr Ragnar Sohlman, who will presumably have to devote most time to this matter, One Hundred Thousand Crowns, and to Mr Rudolf Lilljequist, Fifty Thousand Crowns;

At the present time, my property consists in part of real estate in Paris and San Remo, and in part of securities deposited as follows: with The Union Bank of Scotland Ltd in Glasgow and London, Le Crédit Lyonnais, Comptoir National d'Escompte, and with Alphen Messin & Co. in Paris; with the stockbroker M.V. Peter of Banque Transatlantique, also in Paris; with Direction der Disconto Gesellschaft and Joseph Goldschmidt & Cie, Berlin; with the Russian Central Bank, and with Mr Emanuel Nobel in Petersburg; with Skandinaviska Kredit Aktiebolaget in Gothenburg and Stockholm, and in my strong-box at 59, Avenue Malakoff, Paris; further to this are accounts receivable, patents, patent fees or so-called royalties etc. in connection with which my Executors will find full information in my papers and books.

This Will and Testament is up to now the only one valid, and revokes all my previous testamentary dispositions, should any such exist after my death.

Finally, it is my express wish that following my death my veins shall be opened, and when this has been done and competent Doctors have confirmed clear signs of death, my remains shall be cremated in a so-called crematorium.

Paris, 27 November, 1895

Alfred Bernhard Nobel

That Mr Alfred Bernhard Nobel, being of sound mind, has of his own free will declared the above to be his last Will and Testament, and that he has signed the same, we have, in his presence and the presence of each other, hereunto subscribed our names as witnesses:

Sigurd Ehrenborg	Thos Nordenfelt
former Lieutenant	Constructor
Paris: 84 Boulevard Haussmann	8, Rue Auber, Paris
R. W. Strehlenert	Leonard Hwass
Civil Engineer	Civil Engineer
4, Passage Caroline	4, Passage Caroline

It is interesting to note that Nobel's family received between them a total of Sw. Kr. 1m., equivalent in today's currency to Sw. Kr. 23m. or approximately Two Million Pounds.

Index

Note: In this index Alfred Nobel is abbreviated to AN